Regulation of Serum Lipids by Physical Exercise

Editor

Eino Hietanen, M.D.

Department of Physiology
University of Turku
Turku, Finland
Acting Professor in Physiology
University of Kuopio
Consultant in Clinical Physiology
University Central Hospital of Kuopio
Kuopio, Finland

CRC Press, Inc.
Boca Raton, Florida

Library of Congress Cataloging in Publication Data
Main entry under title:

Regulation of serum lipids by physical exercise.

Bibliography: p.
Includes index.
1. Blood lipoproteins—Metabolism.
2. Exercise—Physiological aspects.
3. Metabolic regulation. 4. Exercise therapy.
I. Hietanen, Eino, 1947– [DNLM:
1. Exertion. 2. Lipids—Blood. 3. Lipids—
Metabolism. 4. Lipoproteins—Blood.
5. Lipoproteins—Metabolism. QU 85 R344]
QP99.3.L52R43 616.1′05 81-38526
ISBN 0-8493-6330-6 AACR2

Direct all inquiries to CRC Press, Inc., 2000 N.W. 24th Street, Boca Raton, Florida 33431.

International Standard Book Number 0-8493-6330-6

Library of Congress Card Number 81-38526
Printed in the United States

PREFACE

Regular exercise training is known to increase both psychical satisfaction and physical fitness by improving cardiorespiratory performance and muscular capacity. Large retrospective and prospective studies have indicated that physical activity most probably decreases cardiovascular mortality. Recently, numerous studies have shown that atherosclerosis and coronary heart disease are related to lipid disorders. Regular exercise may alter serum lipid and lipoprotein profiles. Whether exercise training prevents coronary heart disease by a direct improvement of capillary perfusion in the heart, by decreasing blood pressure, by stabilizing autonomic nervous system, by increasing fibrinolysis and decreasing blood clotting, or whether increased HDL levels improve the transport of cholesterol from vascular walls or decreased LDL levels lower cholesterol accumulation in the vascular intima, is an unanswered question. Possibly physical exercise has multitudinous effects; when a person develops a life style with a regular physical exertion, also other changes in personal habits might take place simultaneously and possibly the number of risk factors decreases.

The aim of this book is to concentrate on the role of physical training in the regulation of serum lipids in healthy man. Exercise conditioning is common among healthy persons and public information advises people to exercise regularly. On the other hand, lipid disorders are among the most common as risk factors of coronary heart disease. It has even been estimated that, when serum cholesterol concentration is normal, the power of other risk factors would be rather low but when cholesterol is high, the power of other risk factors is also multiplied. Regular exercise might also enhance peripheral removal of cholesterol by increasing high density lipoprotein concentration in serum. The object of this book is to estimate the power of regular exercise and the intensity of exercise in the regulation of plasma lipid and lipoprotein concentrations. The mechanisms which mediate the effects of exercise training on serum lipoproteins have first recently begun to be understood. The regulation of lipid metabolism involves numerous pathways and regulatory systems from various hormones to specific enzymes involving many tissues. Our ultimate goal is to find out the basic mechanisms which mediate the exercise-induced changes in serum lipid profile. Thus, basic biochemistry also is of help in evaluating experimental data in this sense.

Finally, specific features of exercise in certain populations will be dealt with. The lipid disorders may originate from childhood which makes it important to practice preventive methods even at these early ages. Thus, attention will be given to estimate the influence of exercise in children. Although our aim is not to provide any profound therapeutic approaches to treat diseases, a short estimation is given on the possibilities of improving physical fitness of persons with coronary heart disease or metabolic disorders. It is our hope that the reader will find this book useful when judging the means of physical exercise in preventive medicine.

Eino Hietanen

THE EDITOR

Eino Hietanen, M.D., is acting professor in physiology at the University of Kuopio and consultant at the Department of Clinical Physiology at the University Hospital in Kuopio, Finland. He is also conducting research at the Department of Physiology, University of Turku and at the Rehabilitation Research Centre of the Finnish Social Insurance Institution.

He obtained his M.D. degree at the University of Turku in Finland and served his residency in clinical physiology at the Departments of Clinical Physiology in Kuopio and Turku University Hospitals. He has been a postdoctoral fellow at the Institute of Human Nutrition, College of Physicians and Surgeons, Columbia University, New York, 1975–1976 and shorter periods at Karolinska Institutet in Stockholm and University of Ulm in West Germany.

His thesis was in physiology of intestinal hydrolytic enzymes and their regulation by nutritional factors. Dr. Hietanen has also conducted research on the role of nutritional factors in the regulation of hepatic and extrahepatic drug metabolism and on the role of dietary factors in experimental carcinogenesis. Other main research objectives include the development of obesity, the role of lungs in lipid metabolism, and especially the regulation of serum lipid metabolism by training and exercise. He has written about 100 original articles and reviews to various international journals and books.

CONTRIBUTORS

Eino Hietanen, M.D.
Department of Physiology
University of Turku
Turku, Finland

Katriina Kukkonen, M.D.
Kuopio Regional Institute
of Occupational Health
Kuopio, Finland

Esko Länsimies, M.D.
Department of Clinical Physiology
University Central Hospital of Kuopio
Kuopio, Finland

Aapo Lehtonen, M.D.
Department of Medicine
University Central Hospital of Turku
Turku, Finland

Jukka Marniemi, Ph.D.
Rehabilitation Research Centre of the
Social Insurance Institution
Turku, Finland

Rainer Rauramaa, M.D., M.Sc.
Institute of Exercise Medicine
Kuopio, Finland

Matti Uusitupa, M.D.
Department of Medicine
University Central Hospital of Kuopio
Kuopio, Finland

Ilkka Välimäki, M.D., M.Sc.
Department of Pediatrics
University Central Hospital of Turku and
Cardiorespiratory Research Unit
University of Turku
Turku, Finland

Jorma Viikari, M.D.
Department of Medicine
University Central Hospital of Turku
Turku, Finland

Erkki Voutilainen, M.D.
Department of Medicine
University Central Hospital of Kuopio
Kuopio, Finland

Veli Ylitalo, M.D.
Department of Pediatrics
University Central Hospital of Turku
Turku, Finland

TABLE OF CONTENTS

CHARACTERIZATION OF LIPOPROTEINS AND THEIR METABOLISM

EXERTION, LIPOPROTEINS AND THEIR METABOLISM

Characterization of Lipoproteins and their Metabolism

Chapter 1

SYNTHESIS AND CATABOLISM

Erkki Voutilainen and Eino Hietanen

TABLE OF CONTENTS

I. FATTY ACIDS AND CHOLESTEROL

The lipoprotein synthesis takes place both in the liver and intestine. Lipid components are synthesized both in the rough and smooth endoplasmic reticulum of cells.[1] The individual apoproteins are probably synthesized in the liver ribosomes. The lipoprotein synthesis is initiated by enzymes catalyzing the esterification of fatty acyl-CoA with glycerol. Fatty acids originate from dietary chylomicrons, liver, and adipose tissue.[1] The relative proportion of each source depends on the composition of the meal and time from the last meal. Fatty acids originating from the adipose tissue are produced by lipolysis catalyzed by the hormone-sensitive lipase which enzyme is activated by cyclic 3′-5′-AMP and by monoacylglycerol lipase.[2–4]

Fatty acid synthetase is stimulated by insulin and fructose and inhibited by glucagon and cyclic AMP.[5,6] Hepatic fatty acids may either be oxidized or synthesized into triglycerides. The fatty acyl-CoA carboxylase enzyme catalyzes the transfer of fatty acids into the mitochondria for the oxidation, while fatty acids exceeding this capacity are partially converted to triglyceride synthesis.[7]

Cholesterol metabolism is another important step in the lipoprotein synthesis, producing one of its main components. The main cholesterol sources are both the dietary cholesterol and endogenous synthesis in the liver and intestine. The net gain of cholesterol is determined by the excretion in bile as such or as neutral sterols and bile acids in feces. A protein binding squalene, called squalene carrier protein(SCP) and 3-hydroxyl-3-methyl-glutaroyl CoA enzyme are among the major determinants in the cholesterol synthesis.[8] SCP resembles apoprotein B yielding to the hypothesis that SCP might serve as a precursor protein for low density lipoprotein (LDL).[8] Human apoprotein A-II present in high density lipoproteins (HDL) has also some similarities to SCP.[9,10]

II. LIPOPROTEINS

The transport of both endogenously synthesized lipids and dietary lipids in the circulation to target organs takes place as lipoprotein complexes. Lipids and proteins together form water-soluble complexes facilitating lipid transport. The classification of plasma lipoproteins is based on their physicochemical characteristics (Table 1).[11–14] Two most common classifications are the motility of lipoproteins in electrophoresis and their gravitational density in ultracentrifugation. Four groups of lipoproteins can be separated with both of these devices. The lipoproteins contain apoproteins, triglycerides, cholesterol, both free and esterified, and phospholipids. The relative proportions of these structural components vary determining, e.g., the gravitational properties of each fraction in the way that the higher density the less lipids this fraction contains. Despite clear classification and characterization of various fractions no clear-cut limits exist for each class but there is a continuous shift from one fraction to the other. Thus the classification to various fractions is partly artificial and serves for analytical purposes. On the other hand, the metabolic behavior of each fraction in the circulation and tissues can be quite different justifying this fractionation.

Even in the same lipoprotein fraction many apoproteins are present. The major apoproteins are apo-A, apo-B, apo-C, and "arginine-rich" apo-E. Apo-A, apo-C, and apo-E have, moreover, subclasses. High density lipoproteins (HDL) contain mainly apo-A peptides, low density lipoproteins (LDL) contain apo-B, and very low density lipoproteins (VLDL) have mainly apo-B, apo-C, and apo-E (Table 1). Practically all apoproteins have been purified and their amino acid sequences have been clarified. The function of apoproteins in lipoproteins is not only structural but they also have specific

Table 1
PHYSICOCHEMICAL PROPERTIES OF LIPOPROTEINS AND THEIR COMPOSITIONAL STRUCTURE[1,11–14]

Lipoprotein class	Ultracentrifugal density (g/ml)	Electrophoretic motility	Lipid composition (% of total weight)				Protein composition								
					Cholesterol			Apoprotein classification (x = minor; xx = major apoprotein)							
			Triglyceride	Phospholipid	Free	Esterified	Total protein	Apo-AI	Apo-AII	Apo-B	Apo-CI	Apo-CII	Apo-CIII	Apo-D	Apo-E
Chylomicron	<0.95	Origin	90	<8	<5		4	x	x	x	xx	xx	xx		x
VLDL	0.95—1.006	Pre-β	50.4—51.5	10.6—11.8	4.8	4.8	8—10			xx	xx	xx	xx	x	x
LDL	1.006—1.063	β	7.5	22.5	7.5	37.5	25—35			xx					
HDL															
HDL$_2$	1.063—1.125	α (HDL$_2$, HDL$_3$)	5	21—26	<15	<7	50	xx	xx					x	x
HDL$_3$	1.125—1.21														
VHDL	>1.21														

metabolic purposes. They may function as enzyme activators or inhibitors and sometimes they may be as specific signals for cell receptors participating in the metabolic regulation (apo-B, apo-E).

Most of the dietary triglycerides are as chylomicrons in the circulation. Chylomicrons are most abundant triglyceride stores and they may contain up to 90% triglycerides (Table 1). Chylomicrons are the least dense lipoproteins due to their high lipid content. Their size is largest, approximately 75 to 600 nm in diameter. The chylomicron structure is formed from triglyceride core surrounded by phospholipids, cholesterol, and apoproteins. They are formed postalimentarily in the epithelial cells of the intestine from fatty acids containing more than ten carbon atoms, and from glycerol via the esterification to triglycerides. When the chylomicrons have been built up from these triglycerides, cholesterol, phospholipids, and apoproteins in the intestinal epithelial cells, they will be secreted to the lymph circulation and further, via the thoracic duct, to the circulation where apo-C from HDL is also incorporated in their structure. The half-life of chylomicrons is rather short, less than 1 hr.[1]

The catabolism of chylomicrons takes place in the peripheral muscle and adipose tissues. The endothelial lipoprotein lipase enzyme hydrolyzes triglycerides to fatty acids and glycerol to be transported to the adipose tissue for storage or to the muscle cells for immediate energy needs. The chylomicron remnants lose part of their apoproteins to HDL and the rest of the remnants are transported to the liver for further catabolism.[1]

Very low density lipoproteins (VLDL) contain mainly endogenous triglycerides synthesized in the liver. Triglycerides form about 50% of VLDL composition while the protein content is about 10% (Table 1). The structure of VLDL molecules resembles that of chylomicrons but they are smaller in size, being 30 to 80 nm in diameter. However, the particle size and composition may vary considerably. Very low density lipoproteins are synthesized in the liver and intestinal epithelial cells. The hepatic VLDL transports endogenously synthesized triglycerides whose fatty acids are mainly from circulating free fatty acids. The intestinal VLDL contains both endogenous and dietary fatty acids in its triglyceride moiety.

The catabolism of VLDL is much similar to that of the chylomicrons. Their triglycerides are hydrolyzed in the adipose and muscle tissues where the endothelial lipoprotein lipase is the key regulatory enzyme. As a product of VLDL catabolism is the low density lipoprotein (LDL) fraction. In concert with the hydrolysis of VLDL triglycerides, most of the surface material of VLDL including phospholipids, cholesterol, and proteins is removed.

Low density lipoproteins (LDL) are in quantity the largest of the lipoprotein fractions.[15] Their structure is composed of lipids (77%), mainly of cholesterol and cholesterol esters (Table 1). About 70% of the serum total cholesterol content is in the LDL fraction. Thus, in cases when total cholesterol is increased, it is due mainly to the increase of this lipoprotein class. The apoprotein moiety of LDL is homogenously made up of apo-B. Structurally this lipoprotein is also spherical in the way that the core is composed of hydrophobic cholesterol esters and triglycerides while the surface is made of polar phospholipids, free cholesterol, and apo-B. Although the lipoprotein lipase enzyme is possibly the rate-limiting factor in the catabolism of VLDL, yielding simultaneously LDL, other processes are also involved.[16] Among the most interesting ones is the transport of apoproteins and free cholesterol to HDL. The most important site for the LDL catabolism is in the peripheral tissues. The LDL particles are bound to the receptor in the cell surface whereafter they are incorporated into the cell for the hydrolysis (Figure 1).[17,18] The liberated LDL cholesterol modifies the intracellular cholesterol synthesis by the inhibition of the 3-hydroxy-3-methyl-glutaroyl coenzyme A reductase activity, thus lowering the intracellular cholesterol production. The apoprotein

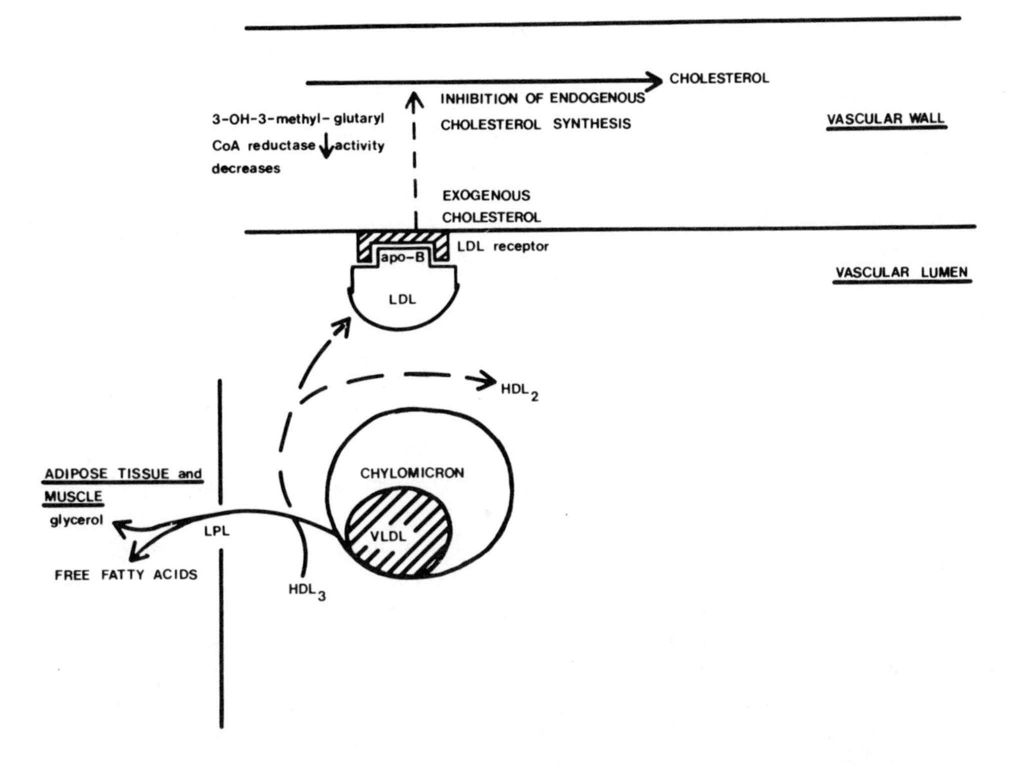

FIGURE 1. Lipoprotein interconversion and catabolism in serum and in participating tissues.

B serves possibly as a signal to the cellular receptors regulating the cholesterol synthesis while a defect in this receptor may prevent the identification of apo-B yielding to the increased intracellular cholesterol accumulation.

The amount of proteins in the heaviest lipoprotein fraction, high density lipoproteins, is about 50%. Most abundant apoproteins are apo-AI and apo-AII while major lipids are phospholipids and cholesterol. The HDL particles are formed in three sites: in the liver, in the intestinal epithelial cells, and as a product of the VLDL triglyceride hydrolysis in the epithelial cells of the vascular endothelium.[19] The difference between the hepatic and intestinal HDL fractions lies in their apoprotein moiety. The hepatic HDL contains both apo-A and apo-C proteins and traces of apo-D and apo-E while that of the intestinal origin contains only apo-A (Table 1).[20] The nascent HDL contains phospholipids, apoproteins, and unesterified cholesterol but is deficient in cholesterol esters. These particles are flat, discoidal, while mature HDL particles are spherical; the core filled with cholesterol esters. The final incorporation of cholesterol esters to the core of nascent HDL is catalyzed by the lecithin cholesterol acyltransferase enzyme.[17,21,22]

The metabolic function of HDL is primarily to transport cholesterol from the peripheral tissues to the liver for the excretion as such, or to be synthesized first to bile acids.[21] The high density lipoproteins have significance also in the metabolism of other lipoproteins as a donor of apoproteins and in the transfer of cholesterol between different lipoprotein fractions. Although the full understanding of the HDL synthesis and catabolism is lacking, some of the main pathways are known.[22–24] Less is known about apoprotein synthesis and about the regulation of apoprotein synthesis. It is possible that factors regulating the apoprotein AI synthesis are specifically of high importance in the overall HDL synthesis. However, better knowledge exists on the enzymatic control of HDL synthesis.[17,23]

III. LIPOLYTIC ENZYMES AND LIPOPROTEINS

Primary invention of the existence of lipoprotein hydrolyzing factor originates back to the 1940s when Hahn[25] found that in dogs intravascular heparin injection postalimentarily cleared the plasma. In the 1950s, Korn[26,27] increased knowledge on this triglyceride lipase, discovering the enzyme in numerous tissues. He further discovered that the activity of this enzyme was dependent on the presence of serum, suggesting that serum contained some components necessary for the enzyme activity. This serum factor was later identified as apoprotein CII which is present in the HDL fraction and is transferred to chylomicrons and VLDL during the lipolysis.[28] The heparin injection liberates the enzyme from tissues into the circulation and the postheparin plasma has consequently been used widely as an enzyme source when the lipoprotein lipase activity has been measured.[29,30]

The lipoprotein lipase enzyme has been found in all species studied and in numerous tissues, such as adipose tissue, muscle, lungs, kidneys, lactating mammary glands, and large blood vessels.[31–33] The functional lipoprotein lipase is located in the capillary endothelium.[34,35] Thus, the enzyme is in close contact with its substrates, chylomicrons, and VLDL triglycerides. The lipoprotein lipase may be bound on the endothelial cell surface to heparan-sulfate which might also explain the liberation of the enzyme to the circulation by heparin injection.[36,37] This can be explained by the higher affinity of the lipoprotein lipase enzyme to heparin than to heparan-sulfate. Although the lipoprotein lipase is located on the endothelial cell surface, it is not synthesized in these cells but in the tissue parenchymal cells.[27,38] In addition to the heparin-releasable extracellular lipoprotein lipase, there is also an intracellular form of lipoprotein lipase, possibly a precursor to the extracellular lipoprotein lipase.[38–40]

The apoprotein CII is an activator of the lipoprotein lipase enzyme.[41–44] Probably apo-CII and lipoprotein lipase form a stable complex in a way that the optimal activity exists at a molar ratio 1:1.[45] It is possible that both the lipoprotein lipase enzyme and apo-CII bind to the lipoprotein surface with their hydrophobic parts.[46,47] The triglyceride hydrolysis products are transported to tissue cells by diffusion for further metabolism.[48] The hydrolysis rate is regulated by the end products through feedback inhibition preventing over-production of end products.[49] Along with the hydrolysis of triglycerides, the composition of chylomicrons and VLDL changes when the triglyceride core becomes smaller as phospholipids, cholesterol, and apo-C are detached from the surface.

The apoprotein CII has affinity to both artificial triglyceride emulsions and lipoprotein lipase enzyme.[45,50] On the other hand, the lipoprotein lipase enzyme is almost completely absorbed to the triglyceride emulsions in the presence of phospholipids, also without apoproteins, but the addition of apo-CII increases the reaction rate.[49] Apo-CII contains both lipid binding and protein binding peptide fragments.[50] To activate the lipoprotein lipase the amino acids of apo-CII from the position 55 to 78 are needed. The peptide fragment with the affinity to lipids is composed of amino acids from 43 to 52 in apo-CII molecule.[50] This follows that apo-CII and the lipoprotein lipase enzyme are bound by their hydrophobic parts to the surface of lipoproteins side by side in the interphase of water-soluble and lipid phases. After the hydrolysis the lipoprotein particle detaches from the endothelial lipoprotein lipase and apo-CII dissociates. Only apo-CII has a peptide sequence able to activate the lipoprotein lipase enzyme; other apo-C proteins inhibit it.[51]

In addition to the lipoprotein lipase, the heparin injection also releases into the circulation the hepatic triglyceride lipase. The postheparin plasma lipolytic activity may even originate for the most part from the liver. However, in enzyme assays the hepatic triglyceride lipase is easily differentiated from the lipoprotein lipase based on different biochemical characteristics.[52] The hepatic lipase has not yet been definitely shown to have any role in the lipoprotein metabolism. However, recent data have suggested that this enzyme might also be involved in the lipoprotein metabolism.[53,54]

REFERENCES

1. **Jackson, R. L., Morrisett, J. D., and Gotto, A. M., Jr.,** Lipoprotein structure and metabolism, *Physiol. Rev.*, 56, 259, 1976.
2. **Khoo, J. C., Aquino, A. A., and Steinberg, D.,** The mechanism of activation of hormone-sensitive lipase in human adipose tissue, *J. Clin. Invest.*, 53, 1124, 1974.
3. **Khoo, J. C., Fong, W. W., and Steinberg, D.,** Activation of hormone-sensitive lipase from human adipose tissue by cyclic AMP-dependent protein kinase, *Biochem. Biophys. Res. Commun.*, 49, 407, 1972.
4. **Tornqvist, H., Nilsson-Ehle, P., and Belfrage, P.,** Enzymes catalyzing the hydrolysis of long-chain monoacylglycerols in rat adipose tissue, *Biochim. Biophys. Acta,* 530, 474, 1978.
5. **Lakshmanan, M. R., Nepokroeff, C. M., and Porter, J. W.,** Control of the synthesis of fatty-acid synthetase in rat liver by insulin, glucagon and adenosine 3′,5′-cyclic monophosphate, *Proc. Natl. Acad. Sci. U.S.A.*, 69, 3516, 1972.
6. **Volpe, J. J. and Vagelos, P. R.,** Regulation of mammalian fatty-acid synthetase. The role of carbohydrate and insulin, *Proc. Natl. Acad. Sci. U.S.A.*, 71, 889, 1974.
7. **McGarry, J. D. and Foster, D. W.,** Regulation of ketogenesis and clinical aspects of the ketotic state, *Metabolism,* 21, 471, 1972.
8. **Scallen, T. J., Srikantaiah, M. V., Skrdlant, H. P., and Hansbury, E.,** Characterization of native sterol carrier protein, *FEBS Lett.*, 25, 227, 1972.
9. **Ritter, M. C. and Dempsey, M. E.,** Specificity and role in cholesterol biosynthesis of a squalene and sterol carrier protein, *J. Biol. Chem.*, 246, 1536, 1971.
10. **Ritter, M. C. and Dempsey, M. E.,** Squalene and sterol carrier protein: structural properties, lipid-binding, and function in cholesterol biosynthesis, *Proc. Natl. Acad. Sci. U.S.A.*, 70, 265, 1973.
11. **Gotto, A. M., Jr., Shepherd, J., Scott, L. W., and Manis, E.,** Primary hyperlipiproteinemia and dietary management, in *Nutrition, Lipids and Coronary Heart Disease,* Levy, R., Rifkind, R., Dennis, B., and Ernst, N., Eds., Raven Press, New York, 1979, 247.
12. **Krehl, W. A.,** The nutritional epidemiology of cardiovascular disease, *Ann. N.Y. Acad. Sci.*, 300, 335, 1977.
13. **Pownall, H. J., Sparrow, J. T., Smith, L. C., and Gotto, A. M., Jr.,** Structure and function of the human plasma apolipoproteins, in *Atherosclerosis,* Vol. 5, Gotto, A. M., Jr., Smith, L. C., and Allen, B., Eds., Springer-Verlag, New York, 1980, 624.
14. **Stein, Y. and Stein, O.,** Metabolism of plasma lipoproteins, in *Atherosclerosis,* Vol. 5, Gotto, A. M., Jr., Smith, L. C., and Allen B., Eds., Springer-Verlag, New York, 1980, 653.
15. **Goldstein, J. L. and Brown, M. S.,** The low-density lipoprotein pathway and its relation to atherosclerosis, *Ann. Rev. Biochem.*, 46, 897, 1977.
16. **Burke, M. D.,** Cholesterol triglyceride and lipoprotein studies: strategies for clinical use, *Postgrad. Med.*, 67, 263, 1980.
17. **Leiss, O., Murawski, U., and Egge, H.,** Lecithin: cholesterol acyltransferase activity in relation to lipoprotein concentration and lipid composition, *Scand. J. Clin. Lab. Invest.*, 38 (Suppl. 150), 77, 1978.
18. **Kinnunen, P. K. J. and Virtanen, L.,** Mode of action of the hepatic endothelial lipase: recycling endocytosis via coated pits, in *Atherosclerosis,* Vol. 5, Gotto, A. M., Jr., Smith, L. C., and Allen, B., Eds., Springer-Verlag, New York, 1980, 383.
19. **Tall, A. R. and Small, D. M.,** Plasma high density lipoproteins, *N. Engl. J. Med.*, 299, 1232, 1978.
20. **Windmueller, H. G. and Spaeth, A. E.,** Fat transport and lymph and plasma lipoprotein biosynthesis by isolated intestine, *J. Lipid Res.*, 13, 92, 1972.
21. **Berger, G. M. B.,** High-density lipoproteins in the prevention of atherosclerotic heart disease. II. Biochemical role in the pathogenesis of atherosclerosis, *S. Afr. Med. J.*, 54, 693, 1978.
22. **Glomset, J. A. and Norum, K. R.,** The metabolic role of lecithin: cholesterol acyltransferase: perspectives from pathology, *Adv. Lipid Res.*, 11, 1, 1973.

23. **Glomset, J. A.,** High density lipoproteins in familial LCAT deficiency, in *High Density Lipoproteins and Atherosclerosis,* Gotto, A. M., Jr., Miller, N. E., and Oliver, M. F., Eds., Elsevier/North-Holland Biomedical Press, Amsterdam, 1978, 57.
24. **Sigurdsson, G., Noel, S.-P., and Havel, R. J.,** Quantification of the hepatic contribution to the catabolism of high density lipoproteins in rats, *J. Lipid Res.,* 20, 316, 1979.
25. **Hahn, P. F.,** Abolishment of alimentary lipemia following injection of heparin, *Science,* 98, 19, 1943.
26. **Korn, E. D.,** Clearing factor, a heparin-activated lipoprotein lipase. I. Isolation and characterization of the enzyme from normal rat heart, *J. Biol. Chem.,* 215, 1, 1955.
27. **Korn, E. D.,** Clearing factor, a heparin-activated lipoprotein lipase. II. Substrate specificity and activation of coconut oil, *J. Biol. Chem.,* 215, 15, 1955.
28. **Fielding, C. J. and Havel, R. J.,** Lipoprotein lipase, *Arch. Pathol. Lab. Med.,* 101, 225, 1977.
29. **Baginsky, M. L. and Brown, W. V.,** Differential characteristics of purified hepatic triglyceride lipase and lipoprotein lipase from human postheparin plasma, *J. Lipid Res.,* 18, 423, 1977.
30. **Bolzano, K.,** Die Bedeutung der Lipoproteinlipase und der hepatischen Triglyzeridlipase beim Abbau Triglyzerid-reicher Lipoproteine, *Wien. Klin. Wochenschr.,* 91, 1, 1979.
31. **Hartiala, J., Viikari, J., Hietanen, E., Toivonen, H., and Uotila, P.,** Cigarette smoke affects lipolytic activity in isolated rat lungs, *Lipids,* 15, 539, 1980.
32. **Hietanen, E. and Hartiala, J.,** Developmental pattern of pulmonary lipoprotein lipase in growing rats, *Biol. Neonate,* 36, 85, 1979.
33. **Rauramaa, R., Kuusela, P., and Hietanen, E.,** Adipose, muscle and lung tissue lipoprotein lipase activities in young streptozotocin treated rats, *Horm. Metab. Res.,* 12, 591, 1980.
34. **Blanchette-Mackie, E. J. and Scow, R. O.,** Sites of lipoprotein lipase activity in adipose tissue perfused with chylomicrons. Electron microscope cytochemical study, *J. Cell. Biol.,* 51, 1, 1971.
35. **Scow, R. O., Blanchette-Mackie, E. J., and Smith, L. C.,** Transport of lipid across capillary endothelium, *Fed. Proc.,* 39, 2610, 1980.
36. **Gertner, S. B. and Sherr, S.,** Quantitative aspects of lipoprotein lipase release by heparin in mice, *Proc. Soc. Exp. Biol Med.,* 162, 389, 1979.
37. **Olivecrona, T., Bengtsson, G., Marklund, S.-E., Lindahl, U., and Höök, M.,** Heparin-lipoprotein lipase interactions, *Fed. Proc.,* 36, 60, 1977.
38. **Nilsson-Ehle, B. and Schotz, M. C.,** A stable, radioactive substrate emulsion for assay of lipoprotein lipase, *J. Lipid Res.,* 17, 536, 1976.
39. **Garfinkel, A. S., Nilsson-Ehle, P., and Schotz, M. C.,** Regulation of lipoprotein lipase induction by insulin, *Biochim. Biophys. Acta,* 424, 264, 1976.
40. **Rodbell, M.,** Metabolism of isolated fat cells, I. Effects of hormones on glucose metabolism and lipolysis, *J. Biol. Chem.,* 239, 375, 1964.
41. **Havel, R. J., Shore, V. G., Shore, B., and Bier, D. M.,** Role of specific glycopeptides of human serum lipoproteins in the activation of lipoprotein lipase, *Circ. Res.,* 27, 595, 1970.
42. **Klose, G., DeGrella, R., Walter, B., and Greten, H.,** Purification of human adipose tissue lipoprotein lipase and human liver lipase, *Artery,* 3, 150, 1977.
43. **Scanu, A.,** Binding of human serum high density lipoprotein apoprotein with aqueous dispersions of phospholipids, *J. Biol. Chem.,* 242, 711, 1967.
44. **Stocks, J. and Galton, D. J.,** Activation of the phospholipase A. Activity of lipoprotein lipase by apoprotein CII, *Lipids,* 15, 186, 1980.
45. **Miller, A. L. and Smith, L. C.,** Activation of lipoprotein lipase by apolipoprotein glutamic acid, *J. Biol. Chem.,* 748, 3359, 1973.
46. **Bensadoun, A. and Kompiang, I. P.,** Role of lipoprotein lipase in plasma triglyceride removal, *Fed. Proc.,* 38, 2622, 1979.
47. **Smith, L. C. and Scow, R. O.,** Intravascular metabolism of lipoproteins: chylomicrons. Mechanisms of transfer of the lipolytic products to cells, in *Progress in Biochemical Pharmacology,* Lipoprotein Metabolism, Vol. 15, Eisenberg, S., Ed., S. Karger, Basel, 1979, 109.
48. **Scow, R. Q., Blanchette-Mackie, E. J., and Smith, L. C.,** Role of capillary endothelium in the clearance of chylomicrons. A model for lipid transport from blood by lateral diffusion in cell membranes, *Circ. Res.,* 39, 149, 1976.
49. **Olivecrona, T. and Bengtsson, G.,** How does lipoprotein lipase bind to substrate lipoproteins and how is its activity regulated?, in *Atherosclerosis,* Vol. 5, Gotto, A. M., Jr., Smith, L. C., and Allen, B., Eds., Springer-Verlag, New York, 1980, 393.
50. **Smith, L. C., Voyta, J. C., Catapano, A. L., Kinnunen, P. K. J., Gotto, A. M., Jr., and Sparrow, J. T.,** Activation of lipoprotein lipase by synthetic fragments of apoC-II, in *Atherosclerosis,* Vol. 5, Gotto, A. M., Jr., Smith, L. C., and Allen, B., Eds., Springer-Verlag, New York, 1980, 397.
51. **Bensadoun, A., Ehnholm, C., Steinberg, D., and Brown, W. V.,** Purification and characterization of lipoprotein lipase from pig adipose tissue, *J. Biol. Chem.,* 249, 2220, 1974.

52. **La Rosa, J. C., Levy, R. I., Windmueller, H. G., and Fredrickson, D. S.,** Comparison of the triglyceride lipase of liver, adipose tissue, and postheparin plasma, *J. Lipid Res.*, 13, 356, 1972.
53. **Jansen, H., van Tol, A., and Hülsmann, W. C.,** On the metabolic function of heparin-releasable liver lipase, *Biochem. Biophys. Res. Commun.*, 92, 53, 1980.
54. **Kuusi, T., Nikkilä, E. A., Virtanen, I., and Kinnunen, P. K. J.,** Localization of the heparin-releasable lipase in situ in the rat liver, *Biochem. J.*, 181, 245, 1979.

Chapter 2

ISOLATION AND DETERMINATION OF LIPOPROTEINS

Erkki Voutilainen and Eino Hietanen

TABLE OF CONTENTS

I. INTRODUCTION

The need for the quantitation of different lipoprotein fractions, instead of total cholesterol and triglyceride determination, is evident according to recent epidemiological, clinical, and experimental studies on the significance of lipoproteins as risk factors—both positive and negative—for atherosclerotic diseases (see Chapter 3). According to their specific structure, the isolation of lipoproteins can be done with ultracentrifugation, selective precipitation, or electrophoretic procedures, and rarely with chromatography. The classification of the lipoproteins, although arbitrary and based on their physicochemical properties, correlates well to their physiological and pathophysiological functions.

Most commonly, lipoproteins are measured as to their cholesterol, triglyceride, or apoprotein contents. Analyses should be done from fresh samples within a few days after collecting them, thus avoiding the deleterious effect of the storage or thawing, particularly on triglyceride-rich lipoproteins.

II. ULTRACENTRIFUGATION

Ultracentrifugation is based on the different density ranges of lipoproteins and other serum proteins.[1–4] So far it has been a reference method for fractionation of lipoprotein classes and subclasses. However, in routine laboratory and clinical practice, it has not been widely used because of its high costs and great need of work.

Sequential preparative ultracentrifugation is the most commonly used technique for separation of serum lipoproteins. Usually, serum samples are spun in an angle-head rotor at +4 to +14° C temperature, adjusting density gradient stepwise with sodium chloride solution, solid sodium, or potassium bromide, according to lipoprotein class or subclass to be fractionated. Running at serum density 1.006 g/ml for 30 min at 105,000 g and exceptionally in a swing-out rotor, chylomicrons flotate to the top fraction and can be separated with Pasteur pipette or preferably by tube slicing for determination. By centrifuging at the same density for 18 hr at 105,000 g, very low density lipoprotein (VLDL) fraction rises to the top of the tube and can be isolated. The subnatant is adjusted to 1.063, centrifugation at 105,000 g for 18 hr is repeated and low density lipoprotein (LDL) class is moved into the top fraction. Further separation of LDL into LDL_1 and LDL_2 can be carried out by spinning the specimen at density 1.019. After isolation of LDL, high density lipoproteins (HDL) remain in the bottom layer. By recentrifuging at 105,000 g for 24 hr after adjusting density to 1.21, other serum proteins sedimentate into the bottom of the tube while HDL floats into supernatant. HDL can also be fractionated at density 1.125 in two subclasses, HDL_2 (1.063 to 1.125) and HDL_3 (1.125 to 1.21). The lipoprotein fractions collected and made up to standard volume are analysed for their different lipid and protein contents.

With the more sophisticated analytical and also with zonal ultracentrifugation,[5–7] the separation and determination of the main lipoprotein classes and subclasses is achieved simultaneously. The recent development of the ultracentrifugal equipment and methods may extend the use of these earlier too expensive and laborious techniques beyond only the research laboratories. Bronzert et al.[8] have described a micromethod using airfuge in the fractionation of lipoproteins and quantitating cholesterol in lipoprotein fractions with an enzymic oxygenelectrode analyzer. By isopycnic density gradient ultracentrifugation, serum lipoproteins have been fractionated by spinning serum samples once for 48 hr.[9]

In the lipoprotein fractionation, conventional preparative ultracentrifugation can be used economically combined with precipitation methods.[3] Triglyceride-rich lipoproteins (chylomicrons, VLDL) are separated with ultracentrifugation, and LDL can be sedimentated from subnatant by some convenient time and cost sparing selective precipitation procedure.

III. PRECIPITATION METHODS

Serum lipoproteins can be isolated selectively and rapidly by precipitation with polyanion and divalent cation according to Burstein et al.,[10–12] either alone or combined with ultracentrifugation. The precipitation of lipoproteins depends upon polyanions and cations used, the concentrations of reagents, pH, ionic strength, and the presence of other serum proteins. By proper choice of these factors, lipoprotein classes can be precipitated stepwise and selectively. The most commonly used polyanions include sulfated polysaccharides like heparin, dextran sulfate, and phosphotungstic acid; the most common divalent cations being magnesium, manganese, and calcium.

Wilson and Spiger[13] and Onononogbu and Lewis[14] have described their precipitation modifications for separating VLDL from LDL and HDL by sodium dodecyl sulfate, thereafter HDL was isolated by dextran sulfate/$CaCl_2$ and LDL was estimated as a difference. A good correlation to results obtained with ultracentrifugation was observed. Simultaneous determination of cholesterol in HDL and LDL fractions is reported by Noma et al.[15] by using heparin and calcium ions as precipitants.

Contributing to recent upsurge interest in HDL, simple polyanion-divalent cation precipitation methods have been widely used on a large scale for isolating HDL-fraction by precipitating apo-B containing lipoproteins (chylomicrons, VLDL, LDL). The common techniques used are the heparin/$MnCl_2$ method,[10–17] the dextran sulfate/$MnCl_2$[10] or the dextran sulfate/$MgCl_2$ procedures,[17–22] and the phosphotungstate/$MgCl_2$ precipitation.[10,13,14,17,21,23] Also, uncharged polymers like polyethylene glycol (PEG) have been used by several authors for serum lipoprotein fractionation.[17,24–27]

The precipitation methods work out in practice as follows. The precipitating reagents, either separately or combined, are added to fresh serum samples and mixed with vortex, incubated at room or +4° C temperature for 5 to 30 min, and centrifuged for 10 to 30 min at 1500 to 12,000 g, varying according to each procedure used. Apo-B containing lipoproteins (chylomicrons, VLDL, and LDL) sedimentate, and HDL can be measured from supernatant. The precipitates of VLDL and LDL can be redissolved and further analyzed.

However, there exist some problems with precipitation procedures either as to the isolation process itself—particularly regarding lipemic samples—or as to the specific and preferable enzymatic cholesterol assay. Warnick et al.[17] have recently evaluated the specificity, accuracy, and precision of commonly used methods.

The heparin/$MnCl_2$ method (Mn^{2+} at 46 mmol/ℓ or at 92 mmol/ℓ) overestimates HDL cholesterol[16,17] owing to incomplete precipitation of apo-B associated lipoprotein classes. So[9] particularly with lipemic samples, ultrafiltering[27] or removal of chylomicrons and VLDL with ultracentrifugation before using this technique is needed. Tallet et al.[28] have also reported complete precipitation of VLDL and LDL with heparin-$MgCl_2$-albumin method from lipemic sera. On the other hand, the precipitative ability of heparin may vary from batch to batch and should be tested before use. Another problem with the heparin/$MnCl_2$ method is its interference with enzymatic cholesterol reagents,[22,29] and the use of EDTA[29] in reagents has not completely eliminated the turbidity difficulties.[30] So high estimation is falsely achieved, and the blank effect of Mn^{2+} ions further raises values.

Better ability to precipitate apo-B containing lipoproteins has been reported with the dextran sulfate/$MgCl_2$,[18,21,22] phosphotungstate/$MgCl_2$,[14,21] and PEG[26,31] techniques.

These methods do not interfere with enzymatic analysis of cholesterol, either. The opinions from the most convenient precipitation method compared to each other are far from clear.[17,21,31] According to our experience, dextran sulfate/$MgCl_2$ procedure, both alone and combined with ultracentrifugation, seems to be the method of choice for the estimation of serum high density lipoprotein cholesterol specifically with automated enzymatic procedure.[22]

At present, the use of the selective precipitation of the lipoproteins is concentrated almost exclusively in isolating antiatherogenic HDL classes. In the future, convenient selective precipitation methods may also be available for isolating other than HDL lipoprotein fraction and subfractions, instead of the laborious methods like ultracentrifugation. For epidemiological purposes, by measuring total cholesterol and triglycerides and HDL cholesterol, LDL can be approximated with Friedewald's formula.[32]

There are obvious systematic differences in HDL levels obtained by different precipitation techniques.[17] Another methodological bias is due to different cholesterol measurement and standardization. Thus it is not easy to compare directly and confidently the risk of atherosclerosis based on data from various methodological studies.

IV. LIPOPROTEIN ELECTROPHORESIS

Electrophoresis has been until recently widely used as a separation method for serum lipoproteins because of the relative simplicity and speed of this technique. Commonly used classification of hyperlipoproteinemias[33] is based on phenotyping lipoproteins electrophoretically. However, the benefit and routine use of qualitative and semiquantitative electrophoresis has recently been considered critically.[34,35]

The charge and molecular diameter of lipoproteins as well as the nature of the medium are affecting the mobility of lipoprotein particles during the electrophoretic run. In modern techniques, paper has been replaced by cellulose acetate,[36] agarose gel,[37,38] and polyacrylamide gel.[39] For detection, lipoproteins have been stained before or after electrophoresis with dyes like Sudan Black and Fat Red 7B. Lately, a more specific enzymatic cholesterol reagent is also used, applying it to the agarose gel[40] or cellulose acetate.[41] Isolated lipoprotein fractions have been quantitated chiefly with scanning densitometry.

The main lipoprotein classes can electrophoretically be classified as chylomicrons, pre-β (VLDL), β (LDL), and α (HDL) fractions according to their typical mobility. Also, with electrophoresis, lipoprotein subgroups can be separated[42,43] and combined with ultracentrifugation. Kremer and Havkes[44] have used electrophoretic assay for determining the ratio of the HDL subgroups.

The use of electrophoresis may be useful in the diagnosis of type III hyperlipoproteinemia, if ultracentrifugation is not available, and also in identifying abnormal lipoprotein states.[45,46] Recently, Muniz[39] and Cobb and Saunders[41] as well as Conlon et al.[40] and Jenny et al.[47] have reported electrophoretic techniques for the measurement of HDL cholesterol. However, Stein et al.[48] preferred the precipitation technique to electrophoresis for separating HDL, in view of the sensitivity, accuracy, and precision needed for the determination of low HDL cholesterol levels.

V. IMMUNOCHEMICAL ASSAYS OF LIPOPROTEINS

The analysis of the apoprotein part is important in fully describing lipoproteins. Physiologically, apoproteins have both catalytic and inhibitory activity in lipoprotein metabolism, so they are not only acting as structural but also as regulatory components (see Chapters 4–6). Clinical experiments, too, show that apoprotein determination is complementary to lipid measurements.[49,50] The immunochemical methods for quantitation of apoproteins from serum or fractionated lipoprotein classes are based on specific antisera for each apoprotein to be determined, and the most commonly used techniques are radial immunodiffusion, electroimmunoassay, and radioimmunoassay.

Radial immunodiffusion[51] is widely used by many authors for determining apo-A and its major constituents A-I and A-II.[52-54] Apo-B, both from serum as well as from LDL and VLDL fractions, can be measured with this technique too.[55,56]

Electroimmunoassay (''rocket'' electrophoresis) for measuring apoproteins in serum and lipoprotein fractions is described by Laurell.[57] Modifications of electroimmunoassay have been used for determination of apolipoprotein A, A-I, and A-II.[58,59] Miller et al.[60] have exposed plasma to urea, instead of delipidation or heating, and were able to detect the full complement of apo-A-I by electroimmunoassay. It has also successfully been used for determination of apo-B,[61,62] and Curry et al.[63] preferred it to other immunochemical methods because of its good precision, accuracy, and particularly speed in measuring apo-B. Minor apoproteins, apo-D[64] and apo-E,[65] have also been determined by electroimmunoassay.

Several authors have described sensitive radioimmunoassays for the quantitation of the main apoprotein A-I of HDL fraction,[66–68] and for apo-B of lower than 1.063 density lipoproteins.[63,69–72]

Also, modifications of nephelometry for estimation of apolipoprotein A-I[73] and apolipoprotein B[74] have been considered recently to offer some advantages like simplicity, rapidity, and to be easily automated as compared to other methods, and to be the method of choice within routine laboratory analysis.

VI. DETERMINATION OF CHOLESTEROL AND TRIGLYCERIDES IN LIPOPROTEINS

The most commonly determined lipid constituents of lipoproteins are cholesterol and triglycerides. For cholesterol determination, colorimetric assays in various modifications[75–78] have been widely used, and so far the methods of Sperry and Webb[75] and of Abell[76] have been used as reference techniques. Nowadays, enzymatic procedures based on cholesterol oxidase[79] coupled with cholesterol esterase,[80] are methods of choice because of their greater specifity compared to chemical methods also measuring nonspecific chromogens. Automated accurate, precise, and specific enzymatic assays are particularly preferable as far as the determination of low cholesterol concentrations in lipoproteins is concerned. Also fluorometric, gas chromatographic, and gravimetric methods have been used for cholesterol quantitation. Triglycerides can be measured chemically by using either colorimetric[81,82] or fluorometric[83,84] determination. Enzymatic manual[85] and automated[86] methods are available, too.

In the future, the promising techniques to be used on a large scale for the isolation of serum lipoproteins and their subclasses include convenient precipitation methods, modifications of immunochemical procedures, and perhaps developed electrophoretic assays. In the category of special ja reference methods remain ultracentrifugation and possibly isoelectric focusing.

REFERENCES

1. **De Lalla, O., and Gofman, J.,** Ultracentrifugal analysis of serum lipoproteins, *Methods Biochem. Anal.*, 1, 459, 1954.
2. **Havel, R. J., Eder, H. A., and Bragdon, J. H.,** The distribution and chemical composition of ultracentrifugally separated lipoproteins in human serum, *J. Clin. Invest.*, 34, 1345, 1955.
3. **Hatch, F. T. and Lees, R. S.,** Practical methods for plasma lipoprotein analysis, *Adv. Lipid Res.*, 6, 1, 1968.
4. **Carlson, K.,** Lipoprotein fractionation, *J. Clin. Pathol.*, 26 (Suppl. 5), 32, 1973.
5. **Charlwood, P. A.,** Density gradient separations in the ultracentrifuge, *Br. Med. Bull.*, 22, 21, 1966.
6. **Wilcox, H. G., Davis, D. C., and Heinberg, M.,** The isolation of lipoproteins from human plasma by ultracentrifugation in zonal rotors, *J. Lipid Res.*, 11, 7, 1970.
7. **Hinton, R. H., Kowalski, A. J., and Mallinson, A.,** Choice of conditions for the gradient flotation of serum lipoproteins in swing-out rotors, *Clin. Chim. Acta*, 44, 267, 1973.
8. **Bronzert, T. J. and Brewer, H. B., Jr.,** New micromethod for measuring cholesterol in plasma lipoprotein fractions, *Clin. Chem.*, 23, 2089, 1977.
9. **Foreman, R. J., Karlin, J. B., Edelstein, C., Juhn, D. J., Rubinstein, A. H., and Scanu, A. M.,** Fractionation of human serum lipoproteins by single-spin gradient ultracentrifugation: quantification of apolipoproteins B and A-I and lipid components, *J. Lipid Res.*, 18, 759, 1977.
10. **Burnstein, M., Scholnick, H. R., and Morfin, R.,** Rapid method for the isolation of lipoproteins from human serum by precipitation with polyanions, *J. Lipid Res.*, 11, 583, 1970.
11. **Burnstein, M. and Scholnick, H. R.,** Lipoprotein-polyanion-metal interactions, *Adv. Lipid Res.*, 11, 67, 1973.
12. **Burstein, M. and Samaille, J.,** Sur un dosage rapide du cholesterol lié aux a- et aux b-lipoprotéines du sérum, *Clin. Chim. Acta*, 5, 609, 1960.
13. **Wilson, D. E. and Spiger, M. J.,** A dual precipitation method for quantitative plasma lipoprotein measurement without ultracentrifugation, *J. Lab. Clin. Med.*, 82, 473, 1973.
14. **Ononogbu, I. C. and Lewis, B.,** Lipoprotein fractionation by a precipitation method. A simple quantitative procedure, *Clin. Chim. Acta*, 71, 397, 1976.
15. **Noma, A., Nezu-Nakayama, K., Kita, M., and Okabe, H.,** Simultaneous determination of serum cholesterol in high- and low-density lipoproteins with use of heparin, Ca^{2+}, and an anion-exchange resin, *Clin. Chem.*, 24, 1504, 1978.
16. **Warnick, G. R. and Albers, J. J.,** A compherensive evaluation of the heparin-manganese precipitation procedure for estimating high density lipoprotein cholesterol, *J. Lipid Res.*, 19, 65, 1978.
17. **Warnick, G. R., Cheung, M. C., and Albers, J. J.,** Comparison of current methods for high-density lipoprotein cholesterol quantitation, *Clin. Chem.*, 25, 596, 1979.
18. **Finley, P. R., Schifman, R. B., Williams, R. J., and Lichti, D. A.,** Cholesterol in high-density lipoprotein: use of Mg^{2+}/dextran sulfate in its enzymic measurment, *Clin. Chem.*, 24, 931, 1978.
19. **Weisweiler, P., Schottdorf, B., and Schwandt, P.,** Cholesterol in high-density lipoproteins: a comparison between dextran-sulphate-magnesium chloride precipitation and preparative ultracentrifugation, *J. Clin. Chem. Clin. Biochem.*, 17, 773, 1979.
20. **Kostner, G. M.,** Enzymatic determination of cholesterol in high-density lipoprotein fractions prepared by polyanion precipitation, *Clin. Chem.*, 22, 695, 1976.
21. **Kostner, G. M., Avogaro, P., Bon, G. B., Cazzolato, G., and Quinci, G. B.,** Determination of high density lipoproteins: screening methods compared, *Clin. Chem.*, 25, 939, 1979.
22. **Penttilä, I. M., Voutilainen, E., Laitinen, P., and Juutilainen, P.,** Comparison of different analytical and precipitation methods for direct estimation of serum high-density lipoprotein cholesterol, *Scand. J. Clin. Lab. Invest.*, in press.
23. **Grove, T. H.,** Effect of reagent pH on determination of high-density lipoprotein cholesterol by precipitation with sodium phosphotungstate-magnesium, *Clin. Chem.*, 25, 560, 1979.
24. **Viikari, J.,** Precipitation of plasma lipoproteins by PEG-6000 and its evaluation with electrophoresis and ultracentrifugation, *Scand. J. Clin. Lab. Invest.*, 36, 265, 1976.
25. **Allen, J. K., Hensley, W. J., Nicholls, A. V., and Whitfield, J. B.,** An enzymatic and centrifugal method for estimating high density lipoprotein cholesterol, *Clin. Chem.*, 25, 325, 1979.
26. **Demacker, P. N. M., Hijmans, A. G. M., Vos-Janssen, H. E., van't Laar, A., and Jansen, A. P.,** A study of the use of polyethylene glycol in estimating cholesterol in high-density lipoprotein, *Clin. Chem.*, 26, 1775, 1980.
27. **Warnick, G. R. and Albers, J. J.,** Heparin-Mn^{2+} quantitation of high-density lipoprotein cholesterol: an ultrafiltration procedure for lipemic samples, *Clin. Chem.*, 24, 900, 1978.
28. **Tallet, F., Raichvarg, D., and Canal, J.,** Heparin-magnesium chloride-albumin method for enzymic measuremnt of cholesterol in high-density lipoprotein, *Clin. Chem.*, 26, 1836, 1980.

29. **Steele, B. W., Koehler, D. F., Azar, M. M., Blászkowski, K. K., and Dempsey, M. E.,** Enzymatic determinations of cholesterol in high-density-lipoprotein fractions prepared by a precipitation technique, *Clin. Chem.*, 22, 98, 1976.
30. **Van der Haar, F., Van Gent, C. M., Schouten, F. M., and Van der Voort, H. A.,** Methods for the estimation of high density cholesterol, comparison between two laboratories, *Clin. Chim. Acta,* 88, 469, 1978.
31. **Demacker, P. N. M., Vos-Janssen, H. E., Hijmans, A. G. M., van't Laar, A., and Jansen, A. P.,** Measurement of high-density lipoprotein cholesterol in serum: comparison of six isolation methods combined with enzymic cholesterol analysis, *Clin. Chem.*, 26, 1780, 1980.
32. **Friedewald, W. T., Levy, R. I., and Fredrickson, D. S.,** Estimation of the concentration of low-density lipoprotein cholesterol in plasma, without use of the preparative ultracentrifuge, *Clin. Chem.*, 18, 499, 1972.
33. **Beaumont, J. H., Carlson, L. A., Cooper, G. R., Fejfar, Z., Fredrickson, D. A., and Strasser, T.,** Classification of the hyperlipidemias and hyperlipoproteinemias, *Bull. W. H. O.*, 43, 891, 1970.
34. **Iammarino, R. M.,** Lipoprotein electrophoresis should be discontinued as a routine procedure, *Clin. Chem.*, 21, 300, 1975.
35. **Fredrickson, D. S.,** It's time to be practical, *Circulation,* 51, 209, 1975.
36. **Chin, H. P. and Blanckenhorn, D. H.,** Separation and quantitative analysis of serum lipoproteins by means of electrophoresis on cellulose acetate, *Clin. Chim. Acta,* 20, 305, 1968.
37. **Noble, R. P.,** Electrophoretic separation of plasma lipoproteins in agarose gel, *J. Lipid Res.*, 9, 693, 1969.
38. **Papadopoulos, N. M. and Kintzios, J. A.,** Determination of serum lipoprotein patterns by agarose gel electrophoresis, *Anal. Biochem.*, 30, 421, 1969.
39. **Muniz, N.,** Measurement of plasma lipoproteins by electrophoresis on polyacrylamide gel, *Clin. Chem.*, 23, 1826, 1977.
40. **Conlon, D. R., Blankstein, L. A., Pasakarnis, P. A., Steinberg, C. M., and D'Amelio, J. E.,** Quantitative determination of high-density lipoprotein cholesterol by agarose gel electrophoresis, *Clin. Chem.*, 25, 1965, 1979.
41. **Cobb, S. A. and Saunders, J. L.,** Enzymatic determination of cholesterol in serum lipoproteins separated by electrophoresis, *Clin. Chem.*, 24, 1116, 1978.
42. **Terebus-Kekish, O., Barclay, M., and Stock, C. C.,** Discrete separation of HDL_2 from HDL_3 of human serum by means of polyacrylamide gel, *Clin. Chim. Acta,* 88, 9, 1978.
43. **Bahler, R. C., Opplt, J. J., and Waggoner, D. M.,** Lipoproteins in patients with proved coronary artery disease; qualitative and quantitative changes in agarose-gel electrophoretic patterns, *Circulation,* 62, 1212, 1980.
44. **Kremer, J. M. H. and Havekes, L.,** Complications in the determination of HDL_2/HDL_3 ratios, *Clin. Chim. Acta,* 109, 21, 1981.
45. **Papadopoulos, N. M.,** Abnormal lipoprotein patterns in human serum as determined by agarose gel electrophoresis, *Clin. Chem.*, 25, 1885, 1979.
46. **Dahlen, G., Ericson, C., Furberg, C., Lunqvist, L., and Svarsudd, K.,** Angina of effort and extra pre-β-lipoprotein fraction, *Acta Med. Scand.*, 531 (Suppl.), 11, 1972.
47. **Jenny, R. W., Newman, H. A. I., and Saat, Y. A.,** A new method for the quantitation of α-lipoprotein cholesterol, *Clin. Chem.*, 24, 1028, 1978.
48. **Stein, E. A., McNeely, S., and Steiner, P.,** Electrophoretic separation of high-density lipoprotein cholesterol evaluated and compared with the modified Lipid Research Clinic Procedure, *Clin. Chem.*, 25, 1934, 1979.
49. **Vergani, C., Trovato, G., and Dioguardi, N.,** Serum total lipids, lipoprotein cholesterol, apoproteins A and B in cardiovascular disease, *Clin. Chim. Acta,* 87, 127, 1978.
50. **Avogaro, P., Cazzolato, G., Bon, G. B., and Quinci, G. B.,** Are apolipoproteins better discriminators than lipids for atherosclerosis?, *Lancet,* 1, 901, 1979.
51. **Mancini, G., Carbonara, A. O., and Heremans, J. F.,** Immunochemical quantitation of antigens by single radial immunodiffusion, *Int. J. Immunochem.*, 2, 235, 1965.
52. **Cheung, M. C. and Albers, J. J.,** The measurement of apolipoprotein A-I and A-II levels in men and women by immunoassay, *J. Clin. Invest.*, 60, 43, 1977.
53. **Albers, J. J., Wahl, P. W., Cabana, V. G., Hazzard, W. R., and Hoover, J. J.,** Quantitation of apolipoprotein A-I of human plasma high-density lipoprotein, *Metabolism,* 25, 633, 1976.
54. **Reman, F. C. and Vermond, A.,** The quantitative determination of apolipoprotein A-I (Apo-lp-GlnI) in human serum by radial immunodiffusion assay (RID), *Clin. Chim. Acta,* 87, 387, 1978.
55. **Lees, R. S.,** Immunoassay of plasma low density lipoproteins, *Science,* 169, 493, 1970.
56. **Kane, J. P., Sata, T., Hamilton, R. L., and Havel, R. J.,** Apoprotein composition of very low density lipoproteins of human serum, *J. Clin. Invest.*, 56, 1622, 1975.

57. **Laurell, C.-B.,** Electroimmunoassay, *Scand. J. Clin. Lab. Invest.*, 29 (Suppl. 124), 21, 1972.
58. **Curry, M. D., Alaupovic, P., and Suenram, C. A.,** Determination of apolipoprotein A and its constitutive A-I and A-II polypeptides by separate electroimmunoassays, *Clin. Chem.*, 22, 315, 1976.
59. **Shepherd, J., Packard, C. J., Patsch, J. R., Gotto, A. M., Jr., and Taunton, O. D.,** Metabolism of apolipoproteins A-I and A-II and its influence on the high density lipoprotein subfraction distribution in males and females, *Eur. J. Clin. Invest.*, 8, 115, 1978.
60. **Miller, J. P., Mao, S. J. T., Patsch, J. R., and Gotto, A. M., Jr.,** The measurement of apolipoprotein A-I in human plasma by electroimmunoassay, *J. Lipid Res.*, 21, 775, 1980.
61. **Kahan, J. and Sunblad, L.,** Immunochemical determination of β-lipoproteins, *Scand. J. Clin. Lab. Invest.*, 24, 61, 1969.
62. **Durrington, P. N., Whicher, J. T., Warren, C., Bolton, C. H., and Hartog, M.,** Comparison of methods for the immunoassay of serum apolipoprotein B in man, *Clin. Chim. Acta*, 71, 95, 1976.
63. **Curry, M. D., Gustafson, A., Alaupovic, P., and McConathy, W. J.,** Electroimmunoassay, radioimmunoassay, and radial immunodiffusion assay evaluated for quantification of human apolipoprotein B., *Clin. Chem.*, 24, 280, 1978.
64. **Curry, M. D., McConathy, W. J., and Alaupovic, P.,** Quantitative determination of human apolipoprotein D by electroimmunoassay and radial immunodiffusion, *Biochim. Biophys. Acta*, 491, 232, 1977.
65. **Curry, M. D., McConathy, W. J., and Alaupovic, P.,** Determination of human apolipoprotein E by electroimmunoassay, *Biochim. Biophys. Acta*, 326, 406, 1976.
66. **Schonfeld, G. and Pfleger, B.,** The structure of human high density lipoprotein and the levels of apolipoprotein A-I in plasma as determined by radioimmunoassay, *J. Clin. Invest.*, 54, 236, 1974.
67. **Karlin, J. B., Juhn, D. J., Starr, J. J., Scanu, A. M., and Rubenstein, A. H.,** Measurement of human high density lipoprotein apoprotein A-I in serum by radioimmunoassay, *J. Lipid Res.*, 17, 30, 1976.
68. **Fainaru, M., Glangeaud, M. C., and Eisenberg, S.,** Radioimmunoassay of human high density lipoprotein A-I, *Biochim. Biophys. Acta*, 386, 432, 1975.
69. **Schonfeld, G., Lees, R. S., George, P. K., and Pfleger, B.,** Assay of total plasma apolipoprotein B concentration in human subjects, *J. Clin. Invest.*, 53, 1458, 1974.
70. **Bautowich, G. J., Simons, L. A., Williams, P. F., and Turtle, J. R.,** Radioimmunoassay of human plasma apolipoproteins, *Atherosclerosis*, 21, 217, 1975.
71. **Thompson, G. P., Birnbaumer, M. E., Levy, R. I., and Gotto, A. M., Jr.,** Solid phase radioimmunoassay of apolipoprotein B (Apo B) in normal human plasma, *Atherosclerosis*, 24, 107, 1976.
72. **Bedford, D. K., Shepherd, J., and Morgan, H. G.,** Radioimmunoassay for human plasma apolipoprotein B, *Clin. Chim. Acta*, 70, 267, 1976.
73. **Shapiro, D., Ballantyne, F. C., and Shepherd, J.,** Comparison of immunonephelometry and electroimmunoassay for estimation of plasma apolipoprotein A-I, *Clin. Chim. Acta*, 103, 7, 1980.
74. **Heuck, C. C. and Schlierf, G.,** Nephelometry of apolipoprotein B in human serum, *Clin. Chem.*, 25, 221, 1979.
75. **Sperry, W. M. and Webb, M.,** A revision of the Schoenheimer-Sperry method for cholesterol determination, *J. Biol. Chem.*, 187, 97, 1950.
76. **Abell, L. L., Levy, B. B., Brodie, B. B., and Kendall, F. E.,** A simplified method for the estimation of total cholesterol in serum and demonstration of its specificity, *J. Biol. Chem.*, 195, 357, 1952.
77. **Mann, G. V.,** A method for measurement of cholesterol in blood serum, *Clin. Chem.*, 7, 275, 1961.
78. **Siegel, A. L. and Bowdoin, B. C.,** Modification of an automated procedure for serum cholesterol which permits the quantitative estimation of cholesterol esters, *Clin. Chem.*, 17, 229, 1973.
79. **Richmond, W.,** Preparation and properties of a cholesterol oxidase from {Nocardia} sp. and its application to the enzymatic assay of total cholesterol in serum, *Clin. Chem.*, 19, 1350, 1973.
80. **Allain, C. C., Poon, L. S., Chan, C. S. G., Richmond, W., and Fu, P. C.,** Enzymatic determination of total serum cholesterol, *Clin. Chem.*, 20, 470, 1974.
81. **Carlson, L. A.,** Determination of serum triglycerides, *J. Atherosclerosis Res.*, 3, 334, 1963.
82. **Laurell, S.,** A method for routine determination of plasma triglycerides, *Scand. J. Clin. Lab. Invest.*, 18, 668, 1966.
83. **Kessler, G. and Lederer, H.,** Fluorometric measurement of triglycerides, in *Automation in Analytical Chemistry*, Skeggs, L. T., Ed., Medical, New York, 1965, 341.
84. **Cramp, D. C. and Robertson, G.,** The fluorometric assay of triglyceride by a semi-automated method, *Anal. Biochem.*, 25, 246, 1968.
85. **Timms, A. R., Kelly, L. A., Spirito, J. A., and Engström, R. G.,** Modification of Hofland's colorimetric semi-automated serum triglyceride determination, assessed by an enzymatic glycerol determination, *J. Lipid Res.*, 9, 675, 1968.
86. **Bell, J. L., Atkinson, M., and Baron, D. N.,** An Auto Analyzer method for estimating serum glyceride glycerol using a glycerokinase procedure, *J. Clin. Pathol.*, 23, 509, 1970.

Chapter 3

LIPIDS AS A RISK FACTOR OF CORONARY HEART DISEASE

Erkki Voutilainen and Eino Hietanen

TABLE OF CONTENTS

I. INTRODUCTION

Much attention has been paid to clarifying the association of serum lipids with the incidence of coronary heart disease. Although no direct cause and consequence relationship between coronary heart disease and serum lipids has been established, numerous epidemiological data have shown a connection between lipids and coronary heart disease. In addition to serum lipids, other major risk factors are cigarette smoking and high blood pressure which when present, increase the coronary heart disease risk five- to tenfold as compared to persons without any of these risk factors.[1-4]

Other factors that increase the incidence of coronary disease include low physical activity, gross obesity, genetic factors, use of oral contraceptives in the presence of one of the other risk factors, and male sex. In this chapter only the lipid classes and their relationship to coronary disease will be discussed.

II. MECHANISMS IN ATHEROSCLEROSIS

The basic mechanisms for the development of atherosclerotic lesions in the vascular walls are far from definite. However, three major theories exist, called lipid theory, blood clotting theory, and mutation theory.[5] The lipid hypothesis is the oldest one. It presumes that atherosclerosis is connected with the accumulation of cholesterol in the vascular walls. This theory is based on both experimental studies showing that dietary manipulation may produce atheromatous lesions in animals and on human studies showing epidemiologically the relationship between the incidence of atheromatous diseases as coronary heart disease and elevated blood lipids. Blood clotting theory is based on the clot formation at the site of the vascular lesion. This is based on the possible lesion of vascular wall intima revealing collagen which induces clot formation at the site of the lesion. The third, and the newest theory, is based on the monoclonal mutation theory suggesting that environmental contaminants and also endogenous reactive intermediates (metabolic products) of, e.g., cholesterol may cause mutation in the myocytes of the vascular wall and start uncontrolled division of these myocytes leading to the plaque formation and slow accumulation of cholesterol later at this site.[6]

Possibly all the presently valid theories in the formation of atheromatous plaque are functionally in harmony. It might well be that the atheromatous plaque begins with the vascular endothelial damage leading to the proliferation of the smooth muscle cells and the production of collagen, elastin, and glycosaminoglycans.[7] Still the initial event leading to the progression of atherosclerotic plaque is unknown. It is not yet definitely established whether the endothelial damage is the initial step in the atheroma formation and how it induces the atheromatous plaque. However, the median smooth muscle cell may play a central role in the arterial cell proliferation and lipid accumulation.[7] Whether low density lipoproteins stimulate the arterial smooth muscle cells in man to proliferate and to accumulate cholesterol esters is still open, although experimental studies do support this.[7] How platelet factors possibly stimulate the growth and proliferation of the smooth muscle cells is also unsolved.

The role of arterial endothelium is to separate the circulating lipoproteins, low density lipoproteins (LDL) and very low density lipoproteins (VLDL), from coming in contact with the media smooth muscle cells and from initiating the events leading to the formation of the atherosclerotic plaque,[6,8] but still being transparent to the necessary nutrients of the cells. The thickness of the endothelial barrier is approximately 320 nm. Essential nutrients are transported through the endothelial membrane by the processes of pinocytosis.[9]

Recent studies have shown that the rate-limiting step in the development of atherosclerotic plaque is the response of the smooth muscle cell to atherosclerosis-promoting factors.[7] It seems that low density lipoproteins play a key role in the transformation of arterial smooth muscle cells to proliferate excessively and to accumulate lipids.[9,10] It might be that the arterial smooth muscle cells have receptors for apoprotein B present in low density lipoproteins, initiating the process of accumulating LDL cholesterol in the arterial wall.[6,7,10] The mechanisms of how the well-known major risk factors are involved in the basic mechanisms of atherosclerosis are far from clear. Both hormonal and mechanical changes related to the hypertension might initiate the endothelial damage in the arteries which is also promoted by the increased carbon monoxide concentration in the blood of cigarette smokers and further due to the damage combined with the increased cholesterol accumulation, especially when serum cholesterol concentration is elevated. The influence of these factors may further be modified by the presence of other promoting factors such as possible autoimmune diseases interacting with the platelet agglutination and viral diseases.[7] Evidence exists that one risk factor alone is not a very potent promoter of atherosclerosis; the endothelial damage does not usually cause atherosclerosis in the absence of increased LDL concentration.[7] This is associated with a low molecular weight polypeptide effect from blood platelets; this polypeptide is essential in the proliferation of the smooth muscle cells of arterial walls.[7]

Despite the accumulation of extracellular material in the vascular wall in atherosclerosis, the disease is basically reversible according to numerous experimental studies and according to epidemiological retrospective and prospective studies in man.[7] In monkeys, experimentally induced atherosclerosis by high cholesterol diets regresses by modifying the diet when measured by the dilatation of the arterial lumen and seen also as a decrease in cholesterol and cholesterol ester contents of vascular walls.[11] Studies with rhesus monkeys have also shown that the regression of experimentally induced atherosclerotic plaque is directly proportional to the decrease of the serum cholesterol concentration.[7] There exist even human studies where the decrease in serum cholesterol concentration has yielded a decrease in the size of atherosclerotic plaque.[7]

The mechanisms on the regression of the plaques are not yet solved. However, some information exists, mainly from experimental studies. The regression includes changes in connective tissue organization in plaques, decrease in cholesterol contents, and healing of the endothelial damage.[7] The decrease in serum cholesterol content is derived from decreased LDL concentration and thus factors promoting lipid accumulation are decreased and the amounts of LDL bound to the media cells are lower, yielding decreased cholesterol accumulation. Simultaneously, often when total cholesterol concentration decreases together with the decrease in LDL concentration, the high density lipoproteins (HDL) concentration increases and at least the relative proportion of the HDL cholesterol vs. total cholesterol is increased. The increase in the proportion of HDL facilitates the removal of peripheral cholesterol and transports cholesterol from the plaque to the liver for further metabolism. Thus, it may well be that the atherosclerotic plaques are not as stable and irreversible as earlier believed but that possibly, e.g., change in the lifestyle may retard the development of atherosclerotic lesions and even promote the regression of already advanced lesions.

III. SERUM CHOLESTEROL AND TRIGLYCERIDE CONCENTRATIONS

Serum total cholestrol concentration has long been acknowledged as one of the major risk factors of atherosclerosis and coronary heart disease. There are also experimental studies supporting the epidemiological data in this respect. The incidence of coronary heart disease varies greatly between countries[6] and also the consumption of dietary fats shows marked geographic variation which has led to studying the connection between dietary lipids and atherosclerosis, especially ischemic heart disease. In epidemiological studies among migrants from countries with low incidence of coronary disease to countries where the coronary heart disease incidence is high, the data have shown that this incidence also increases among men in the new environment with the simultaneous increase in the serum cholesterol concentration.[12] Serum cholesterol concentration is strongly dependent on the dietary fat contents.[13,14] There is strong evidence that the incidence of coronary heart disease is related to the serum total cholesterol concentration.[15,16] In countries with high incidence of coronary heart disease the distribution of serum total cholesterol level is higher than that found among people in countries with a low incidence.[17] The close connection between serum cholesterol concentration and dietary habits is found in international comparative studies which has led to the creation of mathematical formulas to estimate diet induced changes in serum total cholesterol level when dietary habits are changed.[13] In addition to cholesterol, fatty acids also control serum cholesterol levels in the way that saturated fatty acids are twice as powerful in elevating cholesterol concentration as polyunsaturated fatty acids in decreasing it.

Despite the fact that people may live in the same society and have rather similar dietary habits, they differ markedly in their serum cholesterol levels, seen as wide standard deviations.[13] People have been classified in various groups according to their serum total cholesterol level. Although other variables have been standardized, there remains a strong positive association between the risk for the ischemic heart disease and serum total cholesterol level.[3] In attempts to estimate the significance of the cholesterol concentration on the total risk of having coronary disease in a group of 5000 American white males aged 40 to 65 years, about 1000 would develop a coronary heart disease and of these about 300 persons would have it due to their high serum cholesterol level.[18] This calculation, although partly arbitrary, gives a good estimation on the possibilities of controlling one major risk factor to prevent coronary heart disease in a large population. Serum cholesterol concentration over 6.5 mmol/ℓ (250 mg/dℓ) means about a threefold risk of having myocardial infarction in comparison with a person having serum cholesterol below 5.0 mmol/ℓ (194 mg/dℓ).[19,20] Some studies have suggested that of primary importance is the controlling of serum cholesterol levels. When the cholesterol concentration is low, the potential power of cigarette smoking and high blood pressure in initiating coronary disease and other forms of atherosclerosis is lower than in the presence of hypercholesterolemia.

A vast number of studies have been made on the incidence of coronary disease and on the possible risk factors. In the Seven Country Study[17] the prevalence of ischemic heart disease was about five- to sixfold in countries with high prevalence such as eastern Finland and the U.S. as compared to countries with a low prevalence such as southern Europe and Japan.[18] Quite consistently with the incidence of coronary disease, the dietary habits varied while the highest fat consumption was in countries with the high incidence of heart disease and the amount of calories from fats were much lower in countries with the low incidence.[18]

From numerous epidemiological, cross-sectional, and longitudinal studies it is evident that no normal serum cholesterol values can be set up for people in countries with the high consumption of dietary fats in terms that cholesterol values within normal range do not increase the risk of having coronary disease.[18] On the contrary, in most societies with high consumption of dietary fats the goal should be to have as low serum cholesterol concentration as possible.[18,21] In the evaluation of persons on the basis of their serum total cholesterol concentrations it is natural to have the strongest intervention procedures with the highest cholesterol conentrations especially in the cases of "abnormal" levels, i.e., 2 SD above normal mean value. In these cases, drug intervention should also be considered in addition to dietary intervention. In the groups of people with "normal" cholesterol values the physiological intervention methods of decreasing the cholesterol concentration are most valid.

Epidemiological studies have shown that the risk of having coronary heart disease increases continuously when serum cholesterol level exceeds 4.7 mmol/ℓ (180 mg/dℓ) despite the fact that the majority of people in countries on western high fat diets have cholesterol values over 5.7 mmol/ℓ (220 mg/dℓ) which already means a nearly twofold risk of having heart disease in comparison with cholesterol values below 4.7 mmol/ℓ.[22,23] Thus most persons with high serum cholesterol concentration probably do not have a genetic hyperlipoproteinemia but this high level is from environmental factors such as nutritional or other reasons. Also the methods of decreasing these high levels in large populations must be acceptable and suitable for use in large populations, which excludes much of the drug treatment. This group includes a large population who have plasma cholesterol over 6.5 mmol/ℓ (250 mg/dℓ); in the American population about one third is in this category.[24,25]

In vegetarians the serum cholesterol levels are low as in persons with hypobetalipoproteinemia (low LDL levels); both groups have a low incidence of coronary heart disease.[18] In countries where people have low cholesterol levels, persons with coronary disease have elevated cholesterol concentrations in those circumstances where levels might still be normal or even rather low in countries where coronary disease is common and the use of dietary fats high.[18]

Also, in the Stockholm study, total serum triglyceride concentration correlated positively with the incidence of new ischemic heart disease cases in less than 60-years-old men.[26] In a Finnish study by Frick et al.,[27] serum total cholesterol was significantly higher in patients with angiographically found obstruction of coronary arteries even in the way that the more prominent the obstruction the higher the cholesterol concentration. In this study no connection of total triglycerides with coronary heart disease was found, supporting many epidemiological data where no such connection could be established. However, studies suggesting that triglycerides are also involved in the development of atherosclerosis exist. In a study made in Dundee, Scotland, serum lipids were studied in patients admitted to coronary care unit during 1 year.[28] In women the ischemic heart disease correlated with high cholesterol levels and in men with hyperbetalipoproteinemia (LDL). When comparing lipid profiles in men and women having ischemic heart disease with healthy men and women, serum total cholesterol and triglycerides were lower in healthy men than in the ischemic heart disease patients. Also, betalipoproteins were lower in healthy men than in those with the ischemic heart disease.

Although in healthy populations serum total triglyceride levels may not predict the development of coronary heart disease, triglycerides must not be overlooked when judging the lipid profiles.[29] In patients with diabetes, dialysis, or familiar disorders of lipid metabolism the triglyceride concentration may be prognostically significant.

Serum total cholesterol level can be judged as an independent risk factor from other lipid parameters and from other known risk factors. The higher cholesterol the higher risk of having a coronary heart disease even in the way that the high cholesterol level potentiates other risk factors for the coronary heart disease. The total cholesterol concentration has been found to be strongly associated with the coronary heart disease even in the way that the higher cholesterol the more vessels diseased.[30] In this study triglycerides also correlated positively with the existence of coronary disease. Although no conclusive indication exists that high plasma triglyceride is a risk factor of the coronary heart disease, it may prove to be a dependent risk factor related to obesity, blood pressure, and metabolic diseases.[29,31]

IV. CORONARY HEART DISEASE AND LIPOPROTEINS

Fredrickson et al.[32–35] found in their studies the relationship between LDL fraction of lipoproteins and coronary heart disease. Slack[36] found an increased risk of coronary disease in persons with familial hyperlipoproteinemia. Other studies have also confirmed that inherited hyperlipoproteinemic states, either hypercholesterolemia or hypertriglyceridemia, are at least predisposing, albeit not necessarily determining, factors for the development of coronary disease.[37] Gofman et al.[38] were among the first ones to identify the LDL class of lipoproteins elevated in patients with myocardial infarction. Although it has been suggested that the association of LDL with coronary disease is due to the influx of cholesterol from the LDL to arterial wall, many studies have shown far more complicated pathways to exist.[6,10]

In the pursuit of finding possible causes leading to the development of myocardial infarction, long-term compositional changes in serum lipoproteins are of high importance. Studies have been made in postinfarction patients to find out possible long-term changes in the composition of serum lipoproteins in comparison with the healthy controls. Albers et al.[39] studied serum HDL in postinfarction patients when more than 3 months had passed from the infarction. They found, in agreement with studies from persons with coronary artery disease, but not necessarily infarction, that all HDL components, both apoproteins and cholesterol, were lower in postinfarction patients than in healthy lipid-matched controls. In the study by Pometta et al.[40] patients with the myocardial infarction had significantly lower HDL cholesterol concentration than that of healthy controls. Interestingly, male first degree relatives of infarction patients also had significantly lower HDL cholesterol than the control group. In other lipid parameters no equally uniform changes were found in both infarction patients and in their relatives. Also in a study by Kaukola et al.[41] among rather young (mean age 39 years) male survivors of myocardial infarction and age matched controls there was a significant decrease in HDL cholesterol 6 to 20 months after the infarction. An increase in total cholesterol and a decrease in HDL/total cholesterol ratio were also present. Similar results have been drawn from other well-conducted studies.[42–45]

Brunner et al.[46] made a cross-sectional study on serum cholesterol and high-density lipoprotein cholesterol levels in postmyocardial infarction patients and in healthy males and females. In general the HDL cholesterol level was significantly lower in patients with infarction than in healthy controls. Even when the controls with the same cholesterol concentration as those with infarction were studied, the HDL cholesterol concentration was lower in those who had the myocardial infarction. In the study by Brunner et al.,[46] healthy females had higher HDL cholesterol than healthy males, but after the myocardial infarction no difference between males and females was observed. Total cholesterol values were not different in this study between myocardial infarction patients and controls except in the young age group between 35 and 44 years.

Numerous epidemiological studies have been conducted on the relationship between serum lipoproteins and coronary heart disease. Only a few will be cited in this context. In the Tromsø heart study, blood lipids were studied in a relatively young age group of persons (20 to 49 years) during a 2-year case control follow-up.[47] This study was in accordance with others showing a direct relationship between LDL cholesterol and the incidence of coronary disease and an inverse independent relationship between coronary disease and HDL cholesterol. This study suggested that low HDL cholesterol might be a three times stronger predictor for the ischemic heart disease in these relatively young men than high LDL cholesterol concentration.

Castelli et al.[48] studied blood lipids in populations aged over 40 years from different locations and of different ethnic backgrounds. In this study an inverse relationship between HDL cholesterol and coronary heart disease was established in all age groups from 40 to over 70 years, even in the way that the significance of the difference remained in old ages. The HDL cholesterol levels were lowest in persons with ischemic heart disease and although HDL cholesterol was lower in persons who had had the myocardial infarction, it was not as low as in those with coronary disease without infarction. Also in this study, LDL cholesterol and total cholesterol concentrations were elevated in coronary patients when compared to healthy controls. In some coronary disease populations, elevated triglyceride levels were found, but this was not consistent and gave the weakest correlations.[48]

Rhoads et al.[49] studied the relationship between serum lipoproteins and ischemic heart disease in Japanese men aged 50 to 72 years in Hawaii. They found that HDL cholesterol was also an independent inverse risk factor for the coronary disease while total cholesterol and LDL cholesterol levels were positive risk factors. No correlation between the total triglyceride concentration and coronary disease was present. The HDL cholesterol concentration was an independent risk factor from LDL cholesterol, smoking, obesity, and blood pressure.

In a recent epidemiological study on offsprings of the participants of the Framingham Study, the prevalence of coronary heart disease was associated both with total cholesterol and LDL and HDL concentrations as well as total cholesterol/HDL cholesterol ratio.[50] However, VLDL was not related to the prevalence of the coronary disease. In this study HDL cholesterol showed most marked negative correlation with the existence of the coronary disease ($r = -0.618$ to -0.695) while the positive correlation with LDL cholesterol with the multivariate logistic regression analysis was 0.284 to 0.275.[50] Total cholesterol/HDL cholesterol ratios of six or higher were most common in those with the coronary heart disease.

V. HIGH DENSITY LIPOPROTEINS IN RISK EVALUATION

High density lipoprotein carries ordinarily about 20% of the total plasma cholesterol. Although, until recently, major attention has been paid to the total cholesterol, LDL, and VLDL fractions of lipoproteins, Barr et al.[51] found in 1951 that healthy men had higher HDL cholesterol levels than men with coronary heart disease. The same result was also obtained by Nikkilä in 1953.[52] It was already 1966 when Gofman published that especially cholesterol-rich HDL_2 fraction is a good predictor of coronary disease; although first, about 10 years later, HDL cholesterol again gained interest in this respect.[53] In the Cooperative Phenotyping Study[48] a decrease of HDL cholesterol from over 1.17 mmol/ℓ (45 mg/dℓ) to less than 0.65 mmol/ℓ (25 mg/dℓ) increased the coronary disease prevalence from 8 to 19% independently from the plasma triglyceride or LDL cholesterol concentrations. When the data were studied on the basis of ethnic group, sex, or in persons with myocardial infarction or with only coronary disease without infarction, the data always confirmed the lower HDL cholesterol levels in those with coronary disease than in healthy persons. In the Tromsø Heart Study and Framingham Study the predictive power of HDL cholesterol was firmly established.

In the Framingham Study data were analyzed from 2470 men and women over 49 years of age.[54] These data showed that 142 persons had developed a new coronary heart disease during the 2 to 8 years follow-up. The incidence was found to increase 4-fold in men and 12-fold in women when HDL cholesterol value decreased to 38 to 46% of the original level. Quite similar results were obtained in the Tromsø Heart Study.[55] In the Framingham study the power of HDL to predict the development of coronary disease was estimated to be four-fold greater than the LDL cholesterol and eightfold greater than total cholesterol, and in the Tromsø Study HDL had threefold power to LDL cholesterol to predict the coronary heart disease during the 2-year follow-up. In the Israeli Heart Study the HDL cholesterol level was also a risk factor in men having myocardial infarction as well as for the coronary disease without infarction.[56] In this study HDL cholesterol concentration also had prognostic significance when related to sudden deaths and to deaths from myocardial infarction. This Israeli study demonstrated also that the prognostic significance of HDL cholesterol increased during aging. The HDL cholesterol was higher in the group of persons who were occupied by jobs where they had physical activity like walking or physical labor than in those whose work consisted of sitting or standing.

In a London study both men and women having coronary heart disease were found to have lower HDL cholesterol concentration than age-matched controls.[57] When the participants in this study were graded according to their age, a difference between the coronary patients and control persons remained until the age group 60 to 69 years where no difference was found in contrast to some other studies. The vanishing of the difference might be, in part, due to the possible effect of including both sexes.

Wiklund et al.[58] studied HDL composition in patients having myocardial infarction at a young age, below 40 years. They found that apoprotein A was lower in myocardial infarction patients than in age cholesterol, and smoking habits matched controls. This difference was concentrated in the HDL fraction with density 1.063 to 1.21 g/mℓ while in the very high density lipoproteins no difference was present. In this study the HDL triglyceride concentration was higher in the myocardial infarction pateints than in controls, while no difference was found in HDL cholesterol. In the Oslo study the relationship of HDL cholesterol to coronary disease was studied in a fairly young male population aged 40 to 49 years.[59] In this study controls were matched for smoking habits as well as for serum total cholesterol and triglyceride concentrations.[59] Yet in those persons having coronary disease HDL cholesterol was significantly lower (7.9%) than in healthy matched controls and when nonmatched controls were includuded HDL cholesterol was even lower (10.2%). In this study the total cholesterol concentration was significantly lower in both healthy control groups than in coronary patients, and triglycerides were also lower in nonmatched controls.[59] Although in some studies, as in the Israeli Ischemic Heart Disease Study, the predictive power of HDL cholesterol for the coronary heart disease increases with age; in the Olso Study and in many other prospective and retrospective studies data have been obtained suggesting that HDL cholesterol is the most powerful lipid parameter to predict the development of ischemic heart disease at all ages.[48,55]

Of the metabolic inherited disorders of particular interest in regard to HDL fraction of lipoproteins is Tangier's disease, which is characterized by the extremely low HDL concentration in plasma and accumulation of cholesteryl esters in the reticulo-endothelial system. Coronary sclerosis is not increased to the extent anticipated on the relationship between the low HDL cholesterol and incidence of coronary disease in epidemiological studies.[53] However, in addition to the low HDL concentration, the LDL concentration is also extremely low which may contribute to the lower than expected rate of coronary disease. This might suggest that the predictability of low HDL for the development of coronary disease precludes a certain LDL level before valid, although data on this connection is vague. However, confirming data have been found in persons with familial type hyperlipoproteinemia characterized also by low HDL and LDL concentrations.[60] In contrast to the genetically low HDL dyslipoproteinemias, in some genetic disorders a situation where HDL levels are unusually high exists.[61] In these persons the incidence of coronary disease is extremely rare.

The exact explanation why HDL cholesterol has proved to be a better predictor than LDL cholesterol might lie in the unnaturally high LDL levels in western societies. On the basis of the concept of LDL receptor theory, the fairly high LDL levels in most western societies mean saturated LDL receptor sites in the way that an additional increase in the LDL concentration does not occupy any more receptors. On the other hand, a small decrease does not yet liberate any of the receptors due to the oversaturation. Thus factors removing cholesterol from the peripheral tissues become of importance.[62] This might, in part, explain the powerful role of HDL in the prediction of the development of the coronary heart disease.

Williams et al.[63] studied the risk factors of the coronary heart disease and their ratio to the HDL cholesterol level. In this study among healthy British men there was an inverse relationship between the number of risk factors and the level of HDL cholesterol. It has been found by applying sophisticated statistical analysis that, in addition to cholesterol, triglycerides are also a risk factor in certain subgroups. Thus, in the group with the total cholesterol concentration above 6.2 mmol/ℓ (240 mg/dℓ) the high triglyceride level contributes to the total risk although it may be related simultaneously to the decreased HDL levels.[64]

Interestingly, in some societies HDL cholesterol levels are lower in women than in men. One example was found in the British West Indies by Miller and Miller[65] when the average HDL cholesterol value for women was 1.5 mmol/ℓ (59.2 mg/dℓ) and for men 1.72 mmol/ℓ (66.1 mg/dℓ), while this relationship is reversed in North America and in Europe. In this West Indian study women also frequently had ischemic heart disease; their HDL cholesterol, in this study, was also a good predictor for the coronary disease in comparison with LDL which could not separate ischemic persons from healthy ones.

Yet, HDL cholesterol values do not explain all the differences in the incidence of coronary disease in persons of various ethnic backgrounds such as between Japanese and Caucasian Americans or between Scandinavian minorities.[65,66] Further explanations might be found in the more precise analysis of lipoproteins. Marked compositional changes have been found in the HDL fraction due to coronary heart disease in the apoprotein moieties as well as in the cholesterol and phospholipid contents of HDL fraction.[65] Most marked changes take place in HDL_2 fraction in coronary disease while only minor changes are found in the HDL_3 subclass.[65] The cholesterol/protein content of HDL_2 subclass is higher than that of HDL_3. Also the sex difference is mainly due to the difference in HDL_2 concentration, females having higher concentration then males.[65]

From the recent data showing that the low HDL cholesterol content means increased risk for coronary heart disease as do high LDL cholesterol content, it is evident that the significance of total plasma cholesterol in the prediction of coronary disease is less than the significance of whichever two lipoprotein fractions. This favors the partition of lipoproteins in the analysis of blood lipids to estimate the possible risks of having coronary disease or other forms of atherosclerosis.

It has been rather well demonstrated that low concentration of HDL cholesterol precedes the clinical manifestation of coronary heart disease. Studies have also demonstrated decreased apoproteins, apo AI and AII, in coronary disease patients, although probably AI is a less powerful predictor of coronary heart disease than HDL cholesterol.[47] However, Avogaro et al.[67] found that, at least in postinfarction patients, apoproteins might be more predictive than lipoprotein cholesterol.

VI. APOPROTEINS AND ATHEROSCLEROSIS

Onitiri and Jover[68] studied apoproteins in ischemic heart disease. They found marked abnormalities in apoproteins of patients with coronary heart disease. In this study involving 55 patients with ischemic heart disease and 116 healthy controls aged 20 to 70 years of both female and male sexes, marked lipid abnormalities were present among diseased patients.[68] Both total cholesterol and triglyceride concentrations were significantly higher in coronary patients than in controls. In the VLDL class of lipoproteins, both cholesterol and triglyceride concentrations were much higher in the disease group than in controls and the LDL cholesterol was also nearly 30% higher than in the controls. The HDL cholesterol concentration was markedly lower in the diseased group than in controls (1.63 mmol/ℓ vs. 1.43 mmol/ℓ). Even more marked differences were present in the apoproteins. Apoprotein B in VLDL fraction was $2\frac{1}{2}$-fold in coronary patients in comparison with the controls and apo C in this lipoprotein class was over 3-fold in the diseased patients. The apo A decrease in the HDL fraction of coronary patients was only about 25%, being of the same magnitude as the difference in the HDL cholesterol concentration. When the relative amounts of various lipoprotein fractions were calculated, the ratio of apo B to apo C in VLDL was lower in diseased patients than in controls, as was the ratio of VLDL triglyceride to their apoprotein contents in patients. Also, the ratio of LDL cholesterol to their apoproteins was lower in the coronary persons than in controls. On the other hand, HDL cholesterol/apo A ratio remained constant in these groups. The VLDL cholesterol levels showed a negative correlation with HDL cholesterol in both groups suggesting possibly the coupling of HDL formation with the VLDL catabolism. The decreased cleavage of VLDL fraction by the lipoprotein lipase enzyme and decreased enzyme activity have commonly been found in persons with atherosclerosis, possibly relating the lipoprotein lipase enzyme activity to the development of atherosclerosis.[69]

Onitri[70] also compared serum lipoprotein apoproteins in persons with coronary heart disease, peripheral vascular disease, and in healthy controls. The concentrations of apoproteins present in VLDL and LDL fractions of lipoproteins were higher in patients with vascular disease, whether coronary or peripheral, than in controls. The HDL apoproteins were lower in these patients than in controls. Also VLDL triglycerides were highly elevated in both groups of vascular diseases, whereas LDL cholesterol was most markedly elevated in persons with coronary heart disease. However, the concentration of apo B present in LDL fraction was about equally high in both groups of vascular diseases.[70] Also the apo A level was decreased in the HDL fractions of both diseased groups equally, about 25%. It seems that the triglyceride/apoprotein ratio decreases in the VLDL fraction in the vascular atherosclerotic disease. Also cholesterol/apoprotein ratio decreases in LDL fraction indicating that the apoproteins have increased relatively more than cholesterol. On the other hand, in HDL fraction the ratio of HDL cholesterol to apoprotein A remains constant suggesting that both of these lipoprotein components change equally.

The precipitation properties of LDL fraction have been found to be different in patients with coronary heart disease from those of the controls.[71] Postle et al.[72] studied the existence of pre-β peak on agarose gel electrophoresis in myocardial infarction patients, in their relatives, and in healthy controls. Apparently the existence of this pre-β band is coupled with the existence of lipoprotein antigens and is inherited property, although the exact mechanism of inheritance has not been clarified.[73] However, in the study by Postle et al.,[72] the pre-β band was more frequent in patients having myocardial infarction or coronary heart disease than in controls, also being a possible risk factor. In other studies, the existence of intermediate bands between VLDL and LDL fractions in electrophoresis is also common in patients having coronary heart disease.[74]

VII. CORONARY ANGIOGRAPHY AND LIPOPROTEINS

In order to establish the relationship between coronary heart disease and the low HDL cholesterol concentration, Barboriak et al.[75] studied patients who had had coronary angiography and their plasma HDL cholesterol concentrations. In this study patients with symptoms of unstable angina pectoris, stable moderate or severe angina, or those having myocardial infarction previously were included. Some were in examinations due to unknown etiology of chest pain. The lowest occlusion score was in those with the highest HDL cholesterol level. Those with the lowest occlusion score also had the lowest triglyceride conentration, possibly due to the relationship between VLDL triglycerides and HDL cholesterol synthesis catalyzed by the lipoprotein lipase enzyme, although this enzyme was not determined in this study. The alcohol intake was highest in those with the lowest score of the coronary artery occlusion. Similar results in terms of the relationship between HDL cholesterol and coronary artery occlusion were obtained by Tan et al.[76] Other lipid parameters were also changed in the same way as in the study by Barboriak et al.[75] In the study by Tan et al.[76] LDL cholesterol was also elevated in those patients with coronary artery occlusion. When Tan et al.[76] adjusted the HDL cholesterol concentration to the serum triglyceride concentration, the difference in HDL cholesterol between those patients with abnormal coronaries and those with normal coronary arteries but with other signs of coronary disease vanished. However, the difference between HDL cholesterol in healthy persons and those with abnormal coronary angiography remained. Another aspect is that as the HDL formation and VLDL catabolism are related to each other via the lipoprotein lipase enzyme, the adjustment for the triglyceride concentration will vanish any difference in HDL cholesterol due to the fact that VLDL triglycerides containing most circulating triglycerides and HDL cholesterol are dependent on each other inversely.

Pearson et al.[77] found equally convincing data on the connection of low HDL cholesterol level with the advancement of coronary heart disease established by the angiographic study. Wieland et al.[78] found increased total cholesterol and triglyceride levels as well as elevated LDL cholesterol concentration in patients with coronary occlusion. No change was present in HDL cholesterol although HDL/LDL cholesterol ratio was lower in those persons with angiographically confirmed coronary artery disease. Jenkins et al.[79] correlated the degree of coronary atherosclerosis with blood lipid levels. They found that with multivariate analysis there was a positive association in the degree of occlusion with total plasma cholesterol ($r = +0.34$), LDL cholesterol ($r = +0.38$), and LDL cholesterol plus VLDL triglyceride but much stronger negative correlation between HDL cholesterol and coronary artery occlusion ($r = -0.45$).

VIII. CONCLUSIONS

From all of the above and many other studies it is clear that serum lipids and lipoprotein patterns are marked determinants in the development of coronary heart disease and also peripheral artery disease. It is rather well confirmed that high total and LDL cholesterol concentrations mean increased risk of having coronary disease. Recently it has become even more obvious that the low HDL cholesterol concentration, and HDL fraction in general, also means increased risk, and high HDL level has a protecting effect from coronary heart disease. The role of serum triglycerides is less uniformly confirmed while other studies have favored the hypothesis that the high triglyceride level is an independent risk factor and others have shown no relationship between serum total triglyceride concentration and coronary heart disease.[80] However, along with the increase in understanding the mechanisms of the regulation of various lipoproteins, the connection of high triglyceride concentration with the increased risk to have coronary heart disease has become more evident. It seems that the high triglyceride concentration might be only an indirect, dependent risk factor being connected with the HDL metabolism. As HDL is being built up, VLDL triglycerides will be hydrolyzed by the lipoprotein lipase enzyme. When the HDL fraction is low, its formation is slow and consequently the consumption of triglycerides from VLDL fraction is slight, meaning increased triglyceride concentration. While the cumulative evidence is in favor of the proposition that the high HDL fraction prevents coronary heart disease, it is acceptable to find out, at least physiologically, but possibly also by other therapeutical methods, ways to increase the HDL concentration in efforts to decrease the atherosclerosis in its various forms.

REFERENCES

1. **Kannel, W. B., Castelli, W. P., Gordon, T., and McNamara, P. M.,** Serum cholesterol lipoproteins and the risk of coronary heart disease. The Framingham Study, *Ann. Intern. Med.,* 74, 1, 1971.
2. **Reid, D. D., Hamilton, P. J. S., McCartney, P., Rose, G., Jarrett, R. J., and Keen, H.,** Smoking and other risk factors for coronary heart disease in British civil sevants, *Lancet,* 2, 979, 1976.
3. **Stamler, J.,** Dietary and serum lipids in the multifactorial etiology of atherosclerosis, *Arch. Surg.,* 113, 21, 1979.
4. **Wilhelmsen, L., Wedel, H., and Tibbling, G.,** Multivariate analysis of risk factors in coronary heart disease, *Circulation,* 48, 950, 1973.
5. **Olsson, R. E.,** Is there an optimum diet for the prevention of coronary heart disease?, in *Nutrition, Lipids and Coronary Heart Disease,* Levy, R., Rifkind, B., Dennis, B., and Ernst, R., Eds., Raven Press, New York, 1979, 349.
6. **Benditt, E. P. and Gown, A. M.,** Atheroma: the artery wall and the environment, *Int. Rev. Exp. Pathol.,* 21, 55, 1980.
7. **Wissler, R. W.,** Progression and regression of atherosclerotic lesions, *Adv. Exp. Med. Biol.,* 104, 77, 1978.
8. **Leff, D. N.,** Atherosclerosis, *Med. World News,* June 23, 47, 1980.
9. **Scow, R. O., Blanchette-Mackie, E. J., and Smith, L. C.,** Transport of lipid across capillary endothelium, *Fed. Proc. Fed. Soc. Am. Exp. Biol.,* 39, 2610, 1980.
10. **Scanu, A. M.,** Plasma lipoproteins and coronary heart disease, *Ann. Clin. Lab. Sci.,* 8, 79, 1978.
11. **Armstrong, M. L., Warner, E. D., and Connor, W. E.,** Regression of coronary atherosclerosis in rhesus monkeys, *Circ. Res.,* 27, 59, 1970.
12. **Glueck, C. J. and Connor, W. E.,** Diet-coronary heart disease relationships reconnoitered, *Am. J. Clin.,* 31, 727, 1978.
13. **Blackburn, H.,** Diet and mass hyperlipidemia: a public health view, in *Nutrition, Lipids, and Coronary Heart Disease,* Levy, R., Rifkind, B., Dennis, B., and Ernst, N., Eds., Raven Press, New York, 1979, 309.
14. **Turpeinen, O.,** Effect of cholesterol-lowering diet on mortality from coronary heart disease and other causes, *Circulation,* 59, 1, 1979.
15. **Blackburn, H.,** Concepts and controversies about the prevention of coronary heart disease, *Conn. Med.,* 41, 7, 1977.
16. **Blackburn, H.,** Coronary disease prevention. Contoversy and professional attitudes, *Adv. Cardiol.,* 20, 10, 1977.
17. ***Keys, A., Ed.,*** Coronary heart disease in seven countries, *Circulation,* 41 (Suppl. 1), 1970.
18. **Stamler, J.,** Population studies, in *Nutrition, Lipids, and Coronary Heart Disease,* Levy, R., Rifkind, B., Dennis, B., and Ernst, N., Eds., Raven Press, New York, 1979, 25.
19. **Gordon, T., Castelli, W. P., Hjortland, M. C., Kannel, W. B., and Dawber, T. R.,** Predicting coronary heart disease in middle-aged and older persons, *JAMA,* 238, 497, 1977.
20. **Krehl, W. A.,** The nutritional epidemiology of cardiovascular disease, *Ann. N.Y. Acad. Sci.,* 300, 335, 1977.
21. **Margolis, S.,** Physician strategies for the prevention of coronary heart disease, *Johns Hopkins Med. J.,* 141, 170, 1977.
22. **Connor, W. E. and Connor, S. L.,** The key role of nutritional factors in the prevention of coronary disease, *Prev. Med.,* 1, 49, 1972.
23. **Kannel, W. B., Castelli, W. P., Gordon, T., and McNamara, P. M.,** Serum cholesterol, lipoproteins, and the risk of coronary heart disease, *Ann. Intern. Med.,* 74, 1, 1971.

24. **Wood, P. W., Stern, M. P., Silvers, A., Reaven, G. M., and von der Groeben, J.,** Prevalence of plasma lipoprotein abnormalities in a free-living population of the Central Valley, California, *Circulation,* 45, 114, 1972.
25. Levels in Adults, Vital and Health Statistics, U.S. 1960–62, U.S. Public Health Service Publ. 1000, Ser. II, No. 22, National Center for Health Statistics, Washington, D.C., 1967.
26. **Carlson, L. A.,** Plasma lipids and atherosclerosis, *J. Clin. Pathol.,* 26 (Suppl. 5), 43, 1972.
27. **Frick, M. H., Dahlén, G., Berg, K., Valle, M., and Hekali, P.,** Serum lipids in angiographically assessed coronary atherosclerosis, *Chest,* 73, 63, 1978.
28. **Bateson, M. C., Maclean, D., Lowe, K. G., Bouchier, I.A.D., and Evans, J. R.,** Serum lipids and outcome of coronary care unit patients without proven ischaemic heart disease, *Health Bull.,* 36, 220, 1978.
29. **Brunzell, J. D.,** Tirglycerides and coronary heart disease, *N. Engl. J. Med.,* Oct. 30, 1060, 1980.
30. **Cohn, P. F., Gabbay, S. I., and Weglicki, W. B.,** Serum lipid levels in angiographically defined coronary artery disease, *Ann. Intern. Med.,* 84, 241, 1976.
31. **Fuller, J. H., Pinney, S., Jarrett, R. J., Kilbourne, K., and Keen, H.,** Plasma lipids in a London population and their relation to other risk factors for coronary heart disease, *Br. Heart J.,* 40, 170, 1978.
32. **Fredrickson, D. S., Levy, R. I., and Lees, R. S.,** Fat transport in lipoproteins—an integrated approach to mechanism, and disorders, *N. Engl. J. Med.,* 276, 34, 1967.
33. **Fredrickson, D. S., Levy, R. I., and Lees, R. S.,** Fat transport in lipoproteins—an integrated approach to mechanism, and disorders, *N. Engl. J. Med.,* 276, 94, 1967.
34. **Fredrickson, D. S., Levy, R. I., and Lees, R. S.,** Fat transport in lipoproteins—an integrated approach to mechanism, and disorders, *N. Engl. J. Med.,* 276, 214, 1967.
35. **Fredrickson, D. S., Levy, R. I., and Lees, R. S.,** Fat transport in lipoproteins—an integrated approach to mechanism, and disorders, *N. Engl. J. Med.,* 276, 273, 1967.
36. **Slack, J.,** Risk of ischemic heart disease in familial hyperlipoproteinemia states, *Lancet,* 2, 1380, 1969.
37. **Walton, K. W. and Williamson, N.,** Histological and immunofluorescent studies on the evolution of the human atherosclerotic plaque, *J. Atheroscler. Res.,* 8, 599, 1968.
38. **Gofman, J. W., Lindgren, F., Elliott, A., Mantz, W., Hewitt, J., Strisower, B., Herring, B., Herring, V., and Lyon, T. P.,** The role of lipids and lipoproteins in atherosclerosis, *Science,* 111, 166, 1950.
39. **Albers, J. J., Cheung, M. C., and Hazzard, W. R.,** High-density lipoproteins in myocardial infarction survivors, *Metabolism,* 27, 479, 1978.
40. **Pometta, D., Micheli, H., Jornot, C., and Scherrer, J. R.,** HDL-Cholesterol abaissé chez les proches parents et les malades victimes d'infarctus du myocarde, *Schweiz. Med. Wschr.,* 108, 1888, 1978.
41. **Kaukola, S., Manninen, V., and Halonen, P. I.,** Serum lipids with special reference to HDL cholesterol and triglycerides in young survivors of acute myocardial infarction, *Acta Med. Scand.,* 208, 41, 1980.
42. **Chaudhuri, S. and Sundaram, K. R.,** Study of serum lipids and enzymes in myocardial infarction and hypertension, *Thrombosis Res.,* 11, 163, 1977.
43. **Erkelens, D. W., Albers, J. J., Hazzard, W. R., Frederick, R. C., and Bierman, E. L.,** High-density lipoprotein-cholesterol in survivors of myocardial infarction, *JAMA,* 242, 2185, 1979.
44. **Noma, A., Matsushita, S., Komori, T., Abe, K., Okabe, H., Kuramoto, K., and Murakami, M.,** High and low density lipoprotein cholesterol in myocardial and cerebral infarction, *Atherosclerosis,* 32, 327, 1979.
45. **Wiklund, O., Fager, G., Graig, I. H., Wilhelmsson, C.-E., Vedin, A., Olofsson, S.-O., Bondjers, G., and Wilhelmson, L.,** Alphalipoprotein cholesterol levels in relation to acute myocardial infarction and its risk factors, *Scand. J. Clin. Lab. Invest.,* 40, 239, 1980.
46. **Brunner, D., Weisbort, J., Loebl, K., Schwartz, S., Altman, S., Bearman, J. E., and Levin, S.,** Serum cholesterol and high density lipoprotein cholesterol in coronary patients and healthy persons, *Atherosclerosis,* 33, 9, 1979.
47. **Miller, N. E., Förde, O. H., Thelle, D. S., and Mjös, O. D.,** The Tromsø heart study. High-density lipoprotein and coronary heart-disease: a prospective case-control study, *Lancet,* 1, 965, 1977.
48. **Castelli, W. P., Doyle, J. T., Gordon, T., Hames, C. G., Hjortland, M. C., Hulley, S. B., Kagan, A., and Zukel, W. J.,** HDL cholesterol and other lipids in coronary heart disease. The cooperative lipoprotein phenotyping study, *Circulation,* 55, 767, 1977.
49. **Rhoads, G. G., Gulbrandsen, L., and Kagan, A.,** Serum lipoproteins and coronary heart disease in a population study of Hawaii Japanese men, *N. Engl. J. Med.,* 294, 293, 1976.

50. **Wilson, P. W., Garrison, R. J., Castelli, W. P., Feinleib, M., McNamara, P., and Kannel, W. B.,** Prevalence of coronary heart disease in the Framingham Offspring Study: role of lipoprotein cholesterols, *Am. J. Cardiol.*, 46, 649, 1980.
51. **Barr, D. P., Russ, E. M., and Eder, H. A.,** Protein-lipid relationships in human plasma, *Am. J. Med.*, 11, 480, 1951.
52. **Nikkilä, E. A.,** Studies on the lipid-protein relationships in normal and pathological sera and the effect of heparin on serum lipoproteins, *Scand. J. Clin. Lab. Invest.*, 5 (Suppl. 8), 1, 1953.
53. **Miller, N. E.,** The evidence for the artiatherogenicity of high density lipoprotein in man, *Lipids*, 13, 914, 1978.
54. **Gordon, T., Castelli, W. P., Hjortland, M. C., Kannel, W. B., and Dawber, T. R.,** High density lipoprotein as a protective factor against coronary heart disease, The Framingham Study, *Am. J. Med.*, 62, 707, 1977.
55. **Mjøs, O. D., Thelle, D. S., Førde, O. H., and Vik-Mo, H.,** Family study of high density lipoprotein cholesterol and the relation to age and sex. The Tromsø Heart Study, *Acta Med. scand.*, 201, 323, 1977.
56. **Goldbourt, U. and Medalie, J. H.,** High density lipoprotein cholesterol and incidence of coronary heart disease—the Israeli ischemic heart disease study, *Am. J. Epidemiol.*, 109, 296, 1979.
57. **Ononogbu, I. C.,** High density lipoproteins in ischaemic heart disease, *Experientia*, 33, 1063, 1977.
58. **Wiklund, O., Gustafson, A., Bergstrand, R., Vedin, A., and Wilhelmsson, C.,** High density lipoproteins (HDL) in young male myocardial infarction survivors, in *High Density Lipoproteins and Atherosclerosis*, Goth, A. M., Jr., Miller, N. E., and Oliver, M. F., Eds., Elsevier/North-Holland, Amsterdam, 1978, 127.
59. **Enger, S. Chr., Hjermann, I., Foss, O. P., Helgeland, A., Holme, I., Leren, P., and Norum, K. R.,** High density lipoprotein cholesterol and myocardial infarction or sudden coronary death: 'a prospective case-control study in middle-aged men of the Oslo study, *Artery*, 5, 170, 1979.
60. **Fredrickson, D. S. and Levy, R. I.,** Familial hyperlipoproteinemia, in *The Metabolic Bases of Inherited Disease*, Stanbury, J. B., Wyngaarden, J. B., and Fredrickson, D. S., Eds., McGraw-Hill, New York, 1972, 546.
61. **Glueck, C. J., Fallat, R. W., Millet, F., Gartside, P., Elston, R. C., and Go, R. C. P.,** Familial hyper-alpha-lipoproteinemia: studies in eighteen kindreds, *Metabolism*, 24, 1243, 1975.
62. **Havel, R. J.,** High-density lipoproteins, cholesterol transport and coronary heart disease, *Circulation*, 60, 1, 1979.
63. **Williams, P., Robinson, D., and Bailey, A.,** High-density lipoprotein and coronary risk factors in normal men, *Lancet*, 1, 72, 1979.
64. **Scott, D. W., Gorry, G. A., Hoffman, R. G., Barboriak, J. J., and Gotto, A. M.,** A new approach for evaluating risk factors in coronary artery disease: a study of lipid concentrations and severity of disease in 1847 males, *Circulation*, 62, 477, 1980.
65. **Miller, G. J. and Miller, N. E.,** Do high density lipoproteins protect against coronary atherosclerosis?, in *High Density Lipoproteins and atherosclerosis*, Gotto, A. M., Jr., Miller, N. E., and Oliver, M. F., Eds., Elsevier/North-Holland, Amsterdam, 1978, 95.
66. **Førde, O. H., Thelle, D. S., Miller, N. E., and Mjøs, O. D.,** The Tromsø heart study. Distribution of serum cholesterol between high density and lower density lipoproteins in subjects of Norse, Finnish and Lappish ethnic origin, *Acta Med. Scand.*, 203, 21, 1978.
67. **Avogaro, P., Cazzolato, G., Bon, G. B., and Quinci, G. B.,** Are apolipoproteins better discriminators than lipids for atherosclerosis?, *Lancet*, 1, 901, 1979.
68. **Onitiri, A. C. and Jover, E.,** Comparative serum apolipoprotein studies in ischaemic heart disease and control subjects, *Clin. Chim. Acta*, 108, 25, 1980.
69. **Levitova, E. N. and Lobova, N. M.,** Clavage of the fraction of very low density lipoproteins, catalyzed by lipoprotein lipase, from blood sera of normal people and of patients with atherosclerosis, *Vopr. Med. Khim.*, 24, 483, 1978.
70. **Onitri, T. C.,** The possible role of unusually composed lipoproteins in the pathogenesis of atherosclerotic vascular disorders, *Nigerian Med. J.*, 8, 303, 1978.
71. **Waigh, S., Quintero, G., Camejo, G., Acquatella, H., Lalaguna, F., and Bervizbatia, M. L.,** Lipidos sericos y test de precipitation de las betalipoproteinas en pacientes con cardiopatia isquemica y en controles sanos, *Acta Cient. Venezolana*, 28, 89, 1977.
72. **Postle, A. D., Darmady, J. M., and Siggers, D. C.,** Double pre-β-lipoprotein in ischaemic heart disease, *Clin. Gen.*, 13, 233, 1978.
73. **Rhoads, G. G., Morton, N. E., Gulbrandsen, L., and Kagen, A.,** Sinking pre-beta lipoprotein and coronary disease in Japanese-American men in Hawaii, *Am. J. Epidemiol.*, 108, 350, 1978.
74. **Green, J. and Carney, S.,** Molecular exclusion electrophoresis of human serum lipoproteins: patterns in control and ischaemic heart-disease populations, *Clin. Sci. Mol. Med.*, 52, 75, 1977.

75. **Barboriak, J. J., Anderson, A. J., Rimm, A. A., and King, J. F.,** High density lipoprotein cholesterol and coronary artery occlusion, *Metabolism,* 28, 735, 1979.
76. **Tan, M. H., Macintosh, W., Weldon, K. L., Kapoor, A., Chandler, B. M., and Hindmarsh, T. J.,** Serum high density lipoprotein cholesterol in patients with abnormal coronary arteries, *Atherosclerosis,* 37, 187, 1980.
77. **Pearson, T. A., Bulkley, B. H., Achuft, S. C., Kwiterovich, P. O., and Gordis, L.,** The association of low levels of HDL cholesterol and arteriographically defined coronary artery disease, *Am. J. Epidemiol.,* 109, 285, 1979.
78. **Wieland, H., Seidel, D., Wiegand, V., and Kreuzer, H.,** Serum lipoproteins and coronary artery disease (CAD). Comparison of the lipoprotein profile with the results of coronary angiography, *Atherosclerosis,* 36, 269, 1980.
79. **Jenkins, P. J., Harper, R. W., and Nestel, P. J.,** Severity of coronary atherosclerosis related to lipoprotein concentration, *Br. Med. J.,* 2, 388, 1978.
80. **Hulley, S. B., Rosenman, R. H., Bawol, R. D., and Brand, R. J.,** The association between triglyceride and coronary heart disease, *N. Engl. J. Med.,* 302, 1383, 1980.

Chapter 4

REGULATION OF LIPOPROTEIN METABOLISM BY NUTRITIONAL FACTORS

Eino Hietanen

TABLE OF CONTENTS

I. INTRODUCTION

Nutrition is among the most important environmental factors in the regulation of the serum lipoprotein levels. It has long been acknowledged that the total serum cholesterol level is a risk factor for the coronary heart disease.[1] Although the association between the coronary heart disease and dietary factors is hard to establish experimentally due to the different lipoprotein patterns in animals and possibly due to the different mechanisms in the development of atherosclerosis, quite expansive epidemiological studies suggest the relationship between high serum total cholesterol level and coronary heart disease.[2] Later, in more detailed analysis of plasma lipoproteins, a further characterization of serum lipids as risk factors revealed that LDL cholesterol was about equal to the total cholesterol concentration to predict the development of coronary heart disease while HDL cholesterol proved to be an even stronger negative risk factor.[3–5] Respectively, epidemiologically and also in case-control studies, a close connection between low HDL cholesterol and a high risk for coronary heart disease has been found.[3,6–9]

Triglycerides have not proved as important a determinant for the development of coronary heart disease as cholesterol. It may, however, mediate effects through indirect mechanisms, e.g., obesity, diabetes, and hyperlipoproteinemias.[10–12] Although a broad spectrum intervention is always to be applied in the preventive programs for the development of atherosclerosis, nutrition is among the most important ones. Thus, during tens of years, a vast amount of research has been gathered on the role of nutrition in the prevention of the coronary heart disease. Furthermore, numerous studies have been done on the significance of nutritional factors in the regulation of serum lipoproteins (Table 1).[13–17] Despite the large majority of research favoring possibilities of nutritional manipulations to regulate serum lipids and also possibly to influence the development of atherosclerosis, other well-based aspects also exist.[18] Because the main aim of this book is not to give a comprehensive review on the nutritional factors in the regulation of serum lipoproteins and coronary heart disease, only a short summary will be given on nutritional aspects. Moreover, mechanisms will be dealt with on how nutritional factors may mediate their effects on serum lipoproteins. In addition, nutrition may be changed in the course of exercise and training; this possibility should be kept in mind when judging the effects of exercise training on serum lipids.

II. LIPIDS

Animal studies have proposed that dietary cholesterol and saturated fats might have atherogenic effects.[13,19,20] However, experimental animals have quite different lipid profiles in serum from that of man, hampering direct conclusions and applications. The fatty acid composition of dietary triglycerides may be an important determinant in the regulation of serum lipoproteins.[16] Saturated fatty acids seem to increase and polyunsaturated to decrease low density lipoproteins (LDL)[16] although more important than their absolute values is their ratio in the diet.

Just a little is known about the dietary regulation of HDL cholesterol. Feeding extremely high polyunsaturated fat (polyunsaturated: saturated ratio over 4) decreased in a recent study high density lipoproteins (HDL) level and HDL_2: HDL_3 ratio.[21] Other studies do not, however, support this, when the relative amount of polyunsaturated fats were more practicable (polyunsaturated: saturated, 1.5 to 2.0).[22] Whether fats, or their saturation degree, regulate cholesterol synthesis or absorption has not been conclusively evaluated.[23] The response of cholesterol synthesis to dietary fats may vary accordng to species and also in humans individually. The addition of polyunsaturated fat to the diet

Table 1
DIETARY FACTORS AND SERUM LIPOPROTEINS

Diet	HDL	VLDL	LDL	Total triglycerides	Total cholesterol
Cholesterol	↑, ↔	↑	↑		↑
Polyunsaturated fats	↔	↓	↓	↓	↓
Saturated fats		↑	↑		↑
Polyunsaturated/saturated ratio increased	↔	↓	↓		↓
Protein deficiency	↔	↓	↓		↓
Carbohydrate intake	↓	↑		↑	
Ethanol	↑	↔, ↑, ↓	↓, ↔		↔, ↑
Biotin deficiency					↑
Vitamin C excess				↑	
Vitamin D excess					↑
Calcium excess			↓		↓
Dietary fiber (pectins)	↑				↓
Vegetarian diet	↑	↓	↓		↓
Obesity	↓	↑	↑	↑	↔
Weight reduction	↑		↔	↓	↔
Caloric restriction	↓	↔	↑		

Note: ↑ = increase; ↓ = decrease; ↔ = no changes.

increases steroid excretion, possibly suggesting indirectly a slight increase in the cholesterol synthesis in the liver.[23] However, in disorders of the lipid metabolism, the influence of dietary fat saturation may have more pronounced effects.[24] On the other hand, enhanced cholesterol excretion due to the increase of polyunsaturated fats in the diet might mean the increased transit of serum cholesterol to the liver to be excreted and thus yield lowered serum cholesterol levels. The steroid biosynthesis is a large cholesterol consuming process using one third of the daily cholesterol intake. Thus, changes in the bile acid biosynthesis might markedly determine the fate of dietary cholesterol in the body and also the distribution of cholesterol in plasma lipoproteins.

Saturated fatty acids containing triglycerides in the diet increase plasma total cholesterol levels while the diet with high concentration of polyunsaturated fats decreases total cholesterol content.[25–27] The addition of polyunsaturated fats to the diet in the amount of 20% of total energy expediture may result in the decrease of 20 to 30% in the total plasma cholesterol concentration depending on the initial cholesterol level.[23] The main fraction of lipoproteins to be affected by the changes in the saturation degrees of fatty acids is low density lipoproteins (LDL) but very low density lipoproteins (VLDL) fraction also is influenced. Both these fractions are decreased by polyunsaturated fats and increased by saturated fats.

Although the mechanisms of the regulation of plasma lipoproteins by dietary fatty acids have not been completely resolved, data exist favoring both the increased hepatic excretion as well as decreased synthesis.[23] Studies exist suggesting that VLDL synthesis might be decreased due to the high dietary levels of polyunsaturated fats and consequently decrease production of LDL.[28] Suggestions have also been made that the saturation degree of dietary fats might alter the composition of various lipoproteins affecting in this way the amount of cholesterol.[29]

Cholesterol is the most apparent dietary component possibly affecting serum total cholesterol level and cholesterol content in various lipoprotein fractions. However, the evaluation of the dietary cholesterol in the regulation of serum cholesterol levels is very complex and confusing results have been obtained.[18] Apparently the cholesterol source and also the amount of cholesterol related to other dietary lipids might be of importance. Moreover, the effect of cholesterol on serum lipids is not linearly dose-dependent. Excess amounts of dietary cholesterol increase serum total cholesterol usually on an average about 5 to 10%, but individually much higher, and smaller responses have been reported.[16,23,30–32]

Metabolic hyperlipemic states may often be very responsive to dietary manipulation and show marked response to dietary changes.[33] From the dietary point of view Keys et al.[34] have suggested, on the basis of their study, that one egg might increase serum total cholesterol level 0.14 mmol/ℓ or less. Dietary cholesterol has most marked effects on LDL cholesterol, but possibly also high density lipoprotein (HDL), IDL, and VLDL fractions are elevated in response to increases in dietary cholesterol.[23,32] The exact mechanism of how cholesterol will influence the LDL concentration is not resolved. Although no human data exist, animal studies have proposed that the composition of LDL particles and their structure might also change due to dietary cholesterol.[35,36] Possibly dietary cholesterol in man might affect the LDL synthesis in the liver or interact with the VLDL change to LDL in circulation.

In animal studies cholesterol feeding has been found to decrease the normal HDL cholesterol and yield the production of a deviant HDL.[35] In man, however, most studies show no change or increase in HDL cholesterol with high levels of dietary cholesterol; mainly HDL_3 increases and less change is present in HDL_2.[22] In experimental animals a high cholesterol diet may have very drastic effects. In a study by Kushwaha et al.[37] rabbits were fed 0.5% cholesterol diet for 10 to 12 weeks; this feeding pattern resulted in a 40-fold increase in total plasma cholesterol level but no change in the amount of HDL lipoprotein fraction although its protein-cholesterol composition was altered. The addition of cholesterol to the diet either consisting of saturated or polyunsaturated fats yields a similar increase in plasma cholesterol content although the basic level is much lower in persons having a diet rich in polyunsaturated fats.[38]

Although the mechanisms of how dietary lipids control the plasma lipoprotein levels are far from clear, some speculations on the basis of mainly experimental studies can be presented. Studies in laboratory animals suggest that variation in the dietary fats may yield, in addition to quantitative changes in plasma lipoproteins, also qualitative changes.[39] The lipoprotein lipase enzyme participating both in the production of HDL and in the catabolism of VLDL fractions is greatly dependent on dietary lipids. In a study by Cryer et al.[40] they found in guinea pigs that a diet rich in polyunsaturated fats increased markedly the adipose tissue lipoprotein lipase activity (Table 2). This might promote the removal of VLDL triglycerides and possibly facilitate the synthesis of HDL fraction, although in the study no changes in serum lipoproteins were present. Also hepatic fatty acid synthesis might be of importance in lipoprotein metabolism. Generally a high fat diet inhibits while a high carbohydrate diet enhances hepatic fatty acid synthesis as dietary carbohydrates are converted to fatty acids. Lipogenesis is dependent on nutritional and hormonal factors regulating the activities of the acetyl CoA carboxylase and fatty acid synthetase.[41]

In a recent Swedish study the effect of substituting dietary polyunsaturated fats for saturated fats was evaluated in patients with hyperlipoproteinemia.[42] This study was

Table 2
DIET AND LIPOPROTEIN LIPASE ACTIVITY IN ADIPOSE TISSUE AND POSTHEPARIN PLASMA

Decreased activity	Increased activity
Fasting	Refeeding
Caloric restriction	Polyunsaturated fats
Vitamin C deficiency	Carbohydrates
	Obesity

made having patients in metabolic wards. The amount of polyunsaturated fatty acids were increased from the control diet value of 0.2 (ratio polyunsaturated/saturated) to 2.0 in the test diet. Serum total cholesterol content decreased in all hyperlipoproteinemia groups studied (IIa and IIb, IV) by 11 to 13%. The LDL and VLDL cholesterol decreased in all groups and also HDL cholesterol decreased in all groups except in group IV; this was unaltered. Serum total triglyceride content also decreased although the response of triglycerides in individual hyperlipoproteinemia classes was varying in response to polyunsaturated fat diet. Consequently in this study a change in fatty acids was also found in the way that the polyunsaturated-rich diet increased the amount of polyunsaturated fatty acids and decreased saturated fatty acids both in triglycerides, cholesterol esters, and in phospholipids. Another study comparing Eskimos in Greenland and Denmark also suggests that a diet rich in polyunsaturated fats, even of animal origin, may decrease the amount of serum triglycerides and cholesterol.[43] In Greenland Eskimos, the pre-β-lipoprotein levels were much lower than in the Danish counterparts both in men and women, while HDL levels were higher in Greenland men.

III. PROTEINS

When judging the effect of dietary proteins on the plasma lipoproteins, one should consider also, in addition to quantity, the quality and origin of the protein source. In protein deficiency both VLDL and LDL cholesterol fractions are low, while no changes in HDL cholesterol are present.[44] The change of animal protein to plant protein and the simultaneous decrease in quantity has been found to decrease the amount of total serum cholesterol.[45]

In animal studies the change of the diet from animal sources to plant protein, while maintaining the quantity, yielded a highly significant decrease in the amount of total plasma cholesterol.[46] This is also supported by the data indicating that vegetarians have much lower serum cholesterol levels than those eating a mixed diet.[47–49] In type II hyperlipemic patients the substitution of soy protein for animal protein can markedly decrease the serum cholesterol levels.[50] Hill and co-workers[51] studied the effects of dietary factors as possible causes for changes in plasma hormones and lipids in persons at different risks for coronary heart disease. They found that HDL cholesterol was higher in black South African men on a vegetarian diet (their common diet) than in blacks or whites in the U.S. on a normal western diet. When the diets were changed in white North American men, a vegetarian diet decreased plasma triglycerides. When South African blacks were on a western diet, triglycerides and cholesterol increased, HDL and HDL/total cholesterol ratio decreased. The protein influence on serum lipoproteins may also be complicated by the protein-caused changes in the amount of dietary fiber. It is known that a diet of vegetable origin contains larger quantities of fiber than animal diets, possibly preventing the absorption of dietary cholesterol.

IV. CARBOHYDRATES

Carbohydrates have their main effects on VLDL fraction of plasma lipoproteins, thus affecting plasma triglyceride content.[23] The effects of high dietary carbohydrate content are most pronounced in those persons with hypertriglyceridemia.[52] The increase in plasma triglycerides usually precludes, however, that the amount of the dietary carbohydrates is increased to above 65% of the daily calories; this level exceeds that in the average diet. The mechanisms of how excessive dietary carbohydrates increase VLDL triglycerides are partly unsolved.

In experimental animals, dietary sucrose feeding increases triglyceride levels but not cholesterol in serum.[53] The present knowledge suggests that dietary carbohydrates probably promote the triglyceride production in the liver and intestine.[54,55] At the same time a diet rich in carbohydrates, at least in glucose, promotes fatty acid incorporation into the adipocytes by increasing the adipose tissue lipoprotein lipase activity and promoting the development of obesity in experimental animals (Table 2).[56,57] No definite answer can be given as to whether the replacement of dietary fats for carbohydrates decreases cholesterol levels, although some studies supporting this exist.[23] However, possibly greater decrease is obtained in serum cholesterol concentration by substituting polyunsaturated fats for saturated fats.[58] In some studies sucrose and fructose have been considered as more triglyceridemic than glucose.[13] In the evaluation of the significance of carbohydrates in the regulation of serum lipoproteins, the dietary fiber present in foods containing carbohydrates, also may be a modulator of the effect of apparent carbohydrates. At least experimental data show that although carbohydrates as such may not have any effects on the plasma cholesterol levels, the combination of carbohydrates with various lipids may have profound effects.[13] In rabbits, the combination of lactose with cholesterol is highly atherogenic although lactose itself is not.[59]

Carbohydrates have profound effects on plasma HDL levels.[22] Plasma HDL cholesterol level decreases due to the dietary carbohydrate simultaneously with the increase in the VLDL levels.[17] Due to the enhanced catabolism of HDL fraction the HDL apoproteins decrease,[22] also changing the composition of HDL particles.[17] The relative proportion of HDL_3 to HDL_2 increases.[17]

Acute ethanol intake is known to increase plasma triglycerides by increasing the secretion of VLDL lipoproteins from the liver and intestine and by enhancing the fatty acid incorporation from plasma to triglycerides.[60] A slight increase in serum VLDL cholesterol may also result from acute ethanol intake.[61] However, the chronic ethanol intake has no significant correlation with plasma total triglycerides.[62] The fraction mostly influenced by ethanol intake is the high density lipoprotein. In numerous studies a strong positive association between a chronic moderate ethanol intake and plasma HDL cholesterol has been found.[62–67] There is also a negative correlation between LDL cholesterol and ethanol intake.[62] The association between plasma HDL cholesterol and ethanol intake has been confirmed both in epidemiological studies and in prospective studies in humans.[3,62,67–70] In extreme cases the ethanol-induced increase in HDL cholesterol may be many-fold.[71,72]

Ethanol may acutely inhibit the postheparin plasma lipoprotein lipase enzyme promoting hypertriglyceridemia[73] combined with the effects on triglyceride synthesis and secretion. In chronic use ethanol increases the adipose tissue lipoprotein lipase activity and may thus promote the HDL production.[68,74] Nilsson-Ehle et al.[75] found that ethanol intake decreased carbohydrate-induced lipoprotein lipase activity of adipose tissue and decreased the removal capacity for the circulating triglycerides. This might explain the enhanced triglyceride concentration found after ethanol intake. The hypertriglyceridemia produced by acute ethanol intake is not evident in long-term ethanol use when no changes in very low density lipoproteins are found.[3] However, in some cases, ethanol may reveal the underlying primary hypertriglyceridemia.[76] In the evaluation of ethanol in relation to coronary heart disease, it might have moderately used even a protective role as judged from changes in serum lipoprotein profiles and also from studies where ethanol consumption has been correlated with angiographic data.[77,78]

V. MISCELLANEOUS NUTRITIONAL FACTORS

A. Energy Intake

Not only is the composition of diet of importance in the regulation of plasma lipoprotein concentrations, but also the quantitative aspects of nutrition. Overweight is an important problem caused mostly by excessive caloric intake in relation to caloric consumption. Few studies have shown that overweight is negatively related to plasma HDL cholesterol.[79] Also indexes used to characterize overweight correlate negatively with both cholesterol and apoprotein A I.[79,80] On the other hand, adiposity and relative body weight correlate directly with plasma triglycerides.[81] Increased caloric intake may provoke, especially latent, hypertriglyceridemias.[82] Overweight causes especially an increase in VLDL lipoproteins, the major triglyceride carriers. Some overweight persons are able to maintain rather normal plasma VLDL levels but its turnover may be accelerated.[23] This is possibly related to the increased adipose tissue lipoprotein lipase activity, facilitating the hydrolysis of circulating triglycerides and the incorporation of triglycerides in adipose tissue.[83–87] As related to VLDL metabolism, obesity is also possibly related to the increased LDL levels.[82] Restricted dietary energy intake increased LDL triglyceride levels while HDL cholesterol concentration decreased.[88] Although changes in total triglycerides were not significant, these changes correlated well with the changes in the lipoprotein lipase activity in the adipose tissue.[88] Also during fasting, triglyceride concentration decreases; this decrease is well in accordance with the decreased lipoprotein lipase activity during starvation.[89]

B. Dietary Fiber

The question on the significance of fiber and related dietary substances remains difficult to evaluate due to the complexity of the fiber containing material.[16] Cellulose type fibers probably do not decrease serum cholesterol levels but pectin-like fibers decrease total cholesterol levels in man.[90,91] Also leguminous products have been found to decrease serum cholesterol levels.[92] In a recent study a high fiber diet slightly increased HDL cholesterol levels.[93] For further data on the role of dietary fiber, recent reviews are recommended.[94,95] One possible mechanism of action of dietary fiber on serum cholesterol concentration may be mediated by the property of the fiber to bind bile acids and increase their excretion, thus promoting the use of cholesterol on bile acid synthesis and thus the consumption of plasma cholesterol pools.

C. Vitamins

Nicotinic acid is used in pharmacological doses to treat hypertriglyceridemias but at physiological concentrations no definite effect on plasma lipids has been demonstrated. The biotin deficiency is a rare case characterized by the increased cholesterol concentrations.[16] No definite effect of vitamin C on plasma cholesterol concentration has been demonstrated.[16] Experimentally, vitamin C deficiency has been found to produce hypertriglyceridemia probably by decreasing the lipoprotein lipase activity.[96] Overdose of vitamin D may result in hypercholesterolemia associated with hypercalcemia.[97] Vitamin E, despite its role as a biological antioxidant, has no definite role in the regulation of plasma lipoprotein levels.[16]

D. Minerals and Trace Elements

Calcium intake may decrease plasma total cholesterol (mainly the LDL fraction) concentration.[98] However, calcium present in nutrients has not been shown to have any definite role in the regulation of plasma lipoproteins.[16] Iron deficient anemia may be related to decreased plasma cholesterol levels. The role of other minerals and trace elements in the regulation of plasma lipoproteins is far from resolved and needs further work.[16]

E. Others

Although for a long time coffee drinking has been speculated to be a possible risk factor for coronary heart disease, no evidence for this has been found. Coffee drinking has probably no effect on plasma cholesterol concentration.[99,100] However, coffee drinking combined with cigarette smoking may increase the LDL cholesterol concentration.[99]

VI. DIET AND CORONARY HEART DISEASE

Much emphasis has been paid to the nutritional factors and nutritional modification in the prevention of coronary heart disease.[18,101] It is possible that the significance of nutritional factors has been exaggerated, but still the fact remains that the high total cholesterol concentration in plasma means increased risk for the development of coronary heart disease associated with the high LDL cholesterol level.[1] Recently this has been found to be even more pronounced when connected with a low HDL cholesterol level, and the ratio of HDL/total cholesterol has been estimated to be significant as well in the evaluation of the significance of cholesterol levels.[1,3,102] The higher the proportion of HDL cholesterol to the total cholesterol, the more favorable the value is in terms of preventing coronary heart disease.

Thus, there still remains nutritional needs to lower total plasma cholesterol concentration and, on the other hand, to elevate HDL cholesterol.[101] It is reasonably well shown that dietary manipulation modifies the plasma lipoprotein composition. The decreased cholesterol intake with increased amounts of polyunsaturated fats as compared to saturated fats seems to be advisable. Probably it is prognostically significant to decrease the total fat intake. Fortunately these dietary manipulations do not seem to have any marked effects on HDL cholesterol: at least these nutritional changes do not lower HDL cholesterol concentration. Thus, factors increasing the HDL cholesterol concentration combined with dietary factors decreasing total and LDL cholesterol should be positive in terms of preventing coronary heart disease and atherosclerosis.

REFERENCES

1. **Kannel, V. B., Dawber, T. B., Kagan, A., Revotskic, N., and Stokes, J., III,** Factors of risk in the development of coronary heart disease—six year follow up experience: the Framingham Study, *Ann. Intern. Med.*, 55, 33, 1961.
2. **Dawber, T. R., Moore, F. E., and Mann, V. G., II,** Coronary heart disease in the Framingham Study, *Am. J. Public Health,* 47, 4, 1957.

3. **Castelli, W. P., Doyle, J. T., Gordon, T., Hames, C. G., Hjortland, M. C., Hully, S. B., Kagan, A., and Zukel, W. J.,** HDL cholesterol and other lipids in coronary heart disease. The cooperative lipoprotein phenotyping study, *Circulation,* 55, 767, 1977.
4. **Gordon, T., Castelli, W. P., Hjortland, M. C., Kannel, W. B., and Dawber, T. R.,** High density lipoprotein as a protective factor against coronary heart disease. The Framingham Study, *Am. J. Med.,* 62, 707, 1977.
5. **Morris, J. N., Kagan, A., Pattison, D. C., Gordner, M. J., and Raffle, P. A. B.,** Incidence and prediction of ischaemic heart-disease in London busmen, *Lancet,* 2, 553, 1966.
6. **Enger, S. C., Hjermann, J., Foss, O. P., Helgeland, A., Holme, I., Leren, P., and Norum, K. R.,** High density lipoprotein cholesterol and myocardial infarction or sudden coronary death: a prospective case-control study in middle-aged men of the Oslo Study, *Artery,* 5, 170, 1979.
7. **Jenkins, P. J., Harper, R. W., and Nestel, P. J.,** Severity of coronary atherosclerosis related to lipoprotein concentration, *Br. Med. J.,* 2, 388, 1978.
8. **Kannel, W. B.,** Hypertension, blood lipids and cigarette smoking as co-risk factors for coronary heart disease, *Ann. N.Y. Acad. Sci.,* 304, 128, 1978.
9. **Pearson, T. A., Bulkey, B. H., Achugg, S. C., Kwiterovich, P. O., and Cordis, L.,** The association of low levels of HDL cholesterol and arteriographically defined coronary artery disease, *Am. J. Epidemiol.,* 109, 285, 1979.
10. **Heyden, S.,** The problem with triglycerides, *Nutr. Metab.,* 18, 1, 1975.
11. **Rosenman, R. H., Brand, R. J., Scholtz, R. I., and Friedman, M.,** Multivariate prediction of coronary heart disease during 8.5 year follow-up in the Western Collaborative Group Study, *Am. J. Cardiol.,* 37, 903, 1976.
12. **Wilhelmsen, L., Wedel, H., and Tibblin, G.,** Multivariate analysis of risk factors for coronary heart disease, *Circulation,* 48, 950, 1973.
13. **Kritchevsky, D.,** Diet and atherosclerosis, *Am. J. Pathol.,* 84, 615, 1976.
14. **Kritchevsky, D.,** Diet and cholesteremia, *Lipids,* 12, 49, 1977.
15. **Rifkind, B. M., Goor, R. S., and Levy, R. I.,** Current status of the role of dietary treatment in the prevention and management of coronary heart disease, *Med. Clin. N.A.,* 63, 911, 1979.
16. **Truswell, A. S.,** Diet and plasma lipids—a reappraisal, *Am. J. Clin. Nutr.,* 31, 977, 1978.
17. **Wiztum, J. and Schonfeld, G.,** High density lipoproteins, *Diabetes,* 28, 326, 1979.
18. **Mann, G. V.,** Diet-heart: end of an era, *N. Engl. J. Med.,* 297, 644, 1977.
19. **Kritchevsky, D.,** Role of cholesterol vehicle in experimental atherosclerosis, *Am. J. Clin. Nutr.,* 23, 1105, 1970.
20. **Lambert, G. F., Miller, J. P., Olsen, R. T., and Frost, D. V.,** Hypercholesteremia and atherosclerosis induced in rabbits by purified high fat rations devoid of cholesterol, *Proc. Soc. Exp. Biol. Med.,* 97, 544, 1958.
21. **Shepherd, J., Packard, C. J., Patsch, J. R., Gotto, A. M., Jr., and Taunton, O. D.,** Effects of dietary polyunsaturated and saturated fat on the properties of high density lipoproteins and the metabolism of apolipoprotein A-1, *J. Clin. Invest.,* 61, 1582, 1978.
22. **Lewis, B.,** Effects of diets and drugs, in *High Density Lipoproteins and Atherosclerosis,* Gotto, A. M., Jr., Miller, N. E., and Oliver, M. F., Eds., Elsevier/North-Holland Biomedical Press, Amsterdam, 1978, 143.
23. **Grundy, S. M.,** Dietary fats and sterols, in *Nutrition, Lipids, and Coronary Heart Disease,* Levy, R., Rifkind, B., Dennis, B., and Ernst, N., Eds., Raven Press, New York, 1979, 89.
24. **Grundy, S. M., Ahrens, E. H., Jr., and Davignon, J.,** The interaction of cholesterol absorption and cholesterol synthesis in man, *J. Lipid Res.,* 10, 304, 1969.
25. **Leren, P.,** The effect of plasma cholesterol lowering diet in male survivors of myocardial infarction, *Acta Med. Scand.,* Suppl. 466, 1, 1966.
26. **Nestel, P. J., Havenstein, N., Whyte, H. M., Scott, T. J., and Cook, L. J.,** Lowering of plasma cholesterol and enhanced sterol excretion with the consumption of polyunsaturated ruminant fats., *N. Engl. J. Med.,* 288, 379, 1973.
27. **Nestel, P. J. and Homma, Y.,** Effect of dietary polyunsaturated pork on plasma lipids and sterol excretion in man, *Lipids,* 11, 42, 1976.
28. **Grundy, S. M.,** Effects of polyunsaturated fats on lipid metabolism in patients with hypertriglyceridemia, *J. Clin. Invest.,* 55, 269, 1975.
29. **Spritz, N. and Mishkel, M. A.,** Effects of dietary fats on plasma lipids and lipoproteins: a hypothesis for the lipid-lowering effect of unsaturated fatty acids, *J. Clin. Invest.,* 48, 78, 1969.
30. **Keys, A., Anderson, J. T., and Grande, F.,** Serum cholesterol response to changes in the diet. II. The effect of cholesterol in the diet, *Metabolism,* 14, 759, 1965.
31. **Mattson, F. H., Erickson, B. A., and Kligman, A. M.,** Effect of dietary cholesterol on serum cholesterol in man, *Am. J. Clin. Nutr.,* 25, 589, 1972.

32. **Mistry, P., Nicoll, A., Niehaus, C., Christie, I., Janus, E., and Lewis, B.,** Cholesterol feeding revisited, *Circulation,* 54(Suppl. 11), 178, 1976.
33. **Nestel, P. J. and Poyser, A.,** Changes in cholesterol synthesis and excretion when cholesterol intake is increased, *Metabolism,* 25, 1591, 1976.
34. **Keys, A., Grande, F., and Anderson, J. T.,** Bias and misrepresentation revisited: perspective on saturated fat, *Am. J. Clin Nutr.,* 27, 188, 1974.
35. **Mahley, R. W. and Holcombe, K. S.,** Alterations of the plasma lipoproteins and apoproteins following cholesterol feeding in the rat, *J. Lipid Res.,* 18, 314, 1977.
36. **Rudel, L. L., Pitts, L. L., and Nelson, C. A.,** Characterization of plasma low density lipoproteins of nonhuman primates fed dietary cholesterol, *J. Lipid Res.,* 18, 211, 1977.
37. **Kushwaha, R. S., Hazzard, W. R., and Engblom, J.,** High density lipoprotein metabolism in normolipidemic and cholesterol-fed rabbits, *Biochim. Biophys. Acta,* 530, 132, 1978.
38. **Anderson, J. T., Grande, F., and Keys, A.,** Independence of the effects of cholesterol and degree of saturation of the fat in the diet on serum cholesterol in man, *Am. J. Clin. Nutr.,* 29, 1184, 1976.
39. **Calandra, S., Pasquali-Ronchetti, I., Cherardi, E., Fornieri, C., and Tarugi, P.,** Chemical and morphological changes of rat plasma lipoproteins after a prolonged administration of diets containing olive oil and cholesterol, *Atherosclerosis,* 28, 369, 1977.
40. **Cryer, A., Kirtland, J., Jones, H. M., and Gurr, M. I.,** Lipoprotein lipase activity in the tissues of guinea pigs exposed to different dietary fats from conception to three months of age, *Biochem. J.,* 170, 169, 1978.
41. **Volpe, J. J. and Vagelos, P. R.,** Saturated fatty acid biosynthesis and its regulation, *Ann. Rev. Biochem.,* 42, 21, 1973.
42. **Vessby, B., Gustafsson, I.-B., Boberg, J., Karlström, B., Lithell, H., and Werner, I.,** Substituting polyunsaturated for saturated fat as a single change in a Swedish diet: effects on serum lipoprotein metabolism and glucose tolerance in patients with hyperlipoproteinaemia, *Eur. J. Clin. Invest.,* 10, 193, 1980.
43. **Bang, H. O., Dyerberg, J., and Nielsen, A. B.,** Plasma lipid and lipoprotein pattern in Greenlandic west-coast eskimos, *Lancet,* 1, 1143, 1971.
44. **Truswell, A. S.,** Carbohydrate and lipid metabolism in protein-calorie malnutrition, in *Protein-Calorie Malnutrition,* Olson, R. E., Eds., Academic Press, New York, 1975, 119.
45. **Olson, R. E., Vester, J. W., Gursey, D., Davis, N., and Longman, D.,** The effect of low-protein diets upon serum cholesterol in man, *Am. J. Clin. Nutr.,* 6, 310, 1958.
46. **Kritchevsky, D.,** Dietary interactions, in *Nutrition, Lipids, and Coronary Heart Disease,* Levy, R., Rifkind, B., Dennis, B., and Ernst, N., Eds., Raven Press, New York, 1979, 229.
47. **Hardinge, M. G. and Stare, F. J.,** Nutritional studies of vegetarians. II. Dietary and serum levels of cholesterol, *Am. J. Clin. Nutr.,* 2, 83, 1954.
48. **Ruys, J. and Hickie, J. B.,** Serum cholesterol and triglyceride levels in Australian adolescent vegetarians, *Br. Med. J.,* 2, 87, 1976.
49. **Sacks, F. M., Castelli, W. P., Donner, A., and Kass, E. H.,** Plasma lipids and lipoproteins in vegetarians and controls, *N. Engl. J. Med.,* 292, 1148, 1975.
50. **Sirtori, C. R., Agradi, E., Conti, F., Mantero, O., and Gatti, E.,** Soybean-protein diet in the treatment of type II hyperlipoproteinaemia, *Lancet,* 1, 275, 1977.
51. **Hill P., Wynder, E., Garbaczewski, L., Garnes, H., Walker, A. R. P., and Helman, P.,** Plasma hormones and lipids in men at different risk for coronary heart disease, *Am. J. Clin. Nutr.,* 33, 1010, 1980.
52. **Knittle, J. L. and Ahrens, E. H., Jr.,** Carbohydrate metabolism in two forms of hyperglyceridemia, *J. Clin. Invest.,* 43, 485, 1964.
53. **Høstmark, A. T. and Glattre, E.,** Plasma lipid concentration and lipoprotein distribution in exercising and nonexercising rats fed a high sucrose diet, *Experientia,* 35, 627, 1979.
54. **DenBesten, L., Reyna, R. H., Connor, W. E., and Stegink, L. D.,** The different effects on the serum lipids and fecal steroids of high carbohydrate diets given orally or intravenously, *J. Clin. Invest.,* 52, 1384, 1973.
55. **Quarfordt, S. H., Frank, A., Shames, D. M., Berman, M., and Steinberg, D.,** Very low density lipoprotein triglyceride transport in type IV hyperlipoproteinemia and the effects of carbohydrate-rich diets, *J. Clin. Invest.,* 49, 2281, 1970.
56. **Genuth, S. M.,** Effect of high fat vs. high carbohydrate feeding on the development of obesity in weanling ob/ob mice, *Diabetologia,* 12, 155, 1976.
57. **Nilsson-Ehle, P., Carlström, S., and Belfrage, P.,** Rapid effects on lipoprotein lipase activity in adipose tissue of humans after carbohydrate and lipid intake. Time course and relation to plasma glycerol, triglyceride, and insulin levels, *Scand. J. Clin. Lab. Invest.,* 35, 373, 1975.
58. **Whyte, H. M., Nestel, P. J., and Pryke, E. S.,** Bile acid and cholesterol excretion with carbohydrate-rich diets, *J. Lab. Clin. Med.,* 81, 818, 1973.

59. **Wells, W. W. and Anderson, S. C.,** The increased severity of atherosclerosis in rabbits on a lactose-containing diet, *J. Nutr.*, 68, 541, 1959.
60. **Jones, D. P., Perman, E. S., and Lieber, C. S.,** Free fatty acid turnover and triglyceride metabolism after ethanol ingestion in man, *J. Lab. Clin. Med.*, 66, 493, 1965.
61. **Taskinen, M. R. and Nikkilä, E. A.,** Nocturnal hypertriglyceridemia and hyperinsulinemia following moderate evening intake of alcohol, *Acta Med. Scand.*, 202, 1973, 1977.
62. **Castelli, W. P., Doyle, J. T., Gordon, T., Hames, G. C., Hjortland, M., Hylley, S. B., Kagan, A., and Zukcl, W. J.,** Alcohol and blood lipids, *Lancet*, 2, 153, 1977.
63. **Barboriak, J. J., Anderson, A. J., Rimm, A. A., and Tristani, F. E.,** Alcohol and coronary arteries, *Alcoholism Clin. Exp. Res.*, 3, 29, 1979.
64. **Barboriak, J. J. and Menahen, L. A.,** Alcohol, lipoproteins, and coronary heart disease, *Heart Lung*, 8, 736, 1979.
65. **Henze, K., Bucci, A., Signoretti, P., Menotti, A., and Ricci, G.,** Alcohol intake and coronary risk factors in a population group of Rome, *Nutr. Metab.*, 21(Suppl. 1), 157, 1977.
66. **Hirayama, C., Nosaka, Y., Yamada, S., and Yamanishi, Y.,** Effect of chronic ethanol administration on serum high density lipoprotein cholesterol in rat, *Res. Commun. Chem. Pathol. Pharmacol.*, 26, 563, 1979.
67. **Yano, K., Rhoads, G. G., and Kagan, A.,** Coffee, alcohol and risk of coronary heart disease among Japanese men living in Hawaii, *N. Engl. J. Med.*, 297, 405, 1977.
68. **Belfrage, P., Berg, B., Hagerstrand, I., Nilsson-Ehle, P., Tronquist, H., and Webe. T.** Alterations of lipid metabolism in healthy volunteers during long term ethanol intake, *Eur. J. Clin. Invest.*, 7, 127, 1977.
69. **Berg, B. and Johansson, B. G.,** Prolonged administration of ethanol to young healthy volunteers. III. Effects of parameters of liver function, plasma lipid concentrations and lipoprotein patterns, *Acta Med. Scand.*, 194(Suppl. 552), 13, 1973.
70. **Johansson, B. G. and Laurell, C.-B.,** Disorders of serum alpha-lipoproteins after alcoholic intoxication, *Scand. J. Clin. Lab. Invest.*, 23, 231, 1969.
71. **Johansson, C. G. and Medhus, A.,** Increase in plasma α-lipoproteins in chronic alcoholics after acute abuse, *Acta Med. Scand.*, 195, 273, 1974.
72. **Mishkel, M. A.,** Alcohol-Induced (HDL-) Hypercholesterolaemia, *7th Int. Congr. Diabetics*, (Abstr.), Sydney, May 4–10, 1977, 72.
73. **Nikkilä, E. A., Taskinen, M.-R, and Huttunen, J. K.,** Effect of acute ethanol load on postheparin plasma lipoprotein lipase and hepatic lipase activities and intravenous fat tolerance, *Horm. Metab. Res.*, 10, 220, 1978.
74. **Johnson, O. and Hernell, O.,** Effect of ethanol on the activity of lipoprotein lipase in adipose tissue of male and female rats, *Nutr. Metab.*, 19, 41, 1975.
75. **Nilsson-Ehle, P., Carlström, S., and Belgrage, P.,** Rapid effects of ethanol intake on lipoprotein lipase activity in adipose tissue of fasting subjects, *Lipids*, 13, 433, 1978.
76. **Chait, A., Mancini, M., February, A. W., and Lewis, B.,** Clinical and metabolic study of alcoholic hyperlipidaemia, *Lancet*, 2, 62, 1972.
77. **Barboriak, J. J., Anderson, A. J., and Hoffmann, R.,** Interrelationship between coronary artery occlusion, high-density lipoprotein cholesterol, and alcohol intake, *J. Lab. Clin. Med.*, 94, 348, 1979.
78. **Barboriak, J. J., Andersson, A. J., Rimm, A. A., and King, J. F.,** High density lipoproteins cholesterol and coronary artery occlusion, *Metabolism*, 78, 735, 179.
79. **Wilson, D. E. and Lees, R. S.,** Metabolic relationships among the plasma lipoproteins. Reciprocal changes in the very-low and low-density lipoproteins in man, *J. Clin. Invest.*, 51, 1051, 1972.
80. **Avogaro, P., Cazzolato, G., Bittolo, Bon, G., Guinci, C. B., and Chinello, M.,** HDL-cholesterol, apolipoproteins A_1 and B. Age and index body weight, *Atherosclerosis*, 31, 85, 1978.
81. **Myers, L. H., Phillips, N. R., and Havel, R. J.,** Mathematical evaluation of methods for estimation of the concentration of the major lipid components of human serum lipoproteins, *J. Lab. Clin. Med.*, 88, 491, 1976.
82. **Olefsky, J., Reaven, G. M., and Farquhar, J. W.,** Effects of weight reduction on obesity. Studies of lipid and carbohydrate metabolism in normal and hyperlipoproteinemic subjects, *J. Clin. Invest.*, 53, 64, 1974.
83. **Taskinen, M.-R. and Nikkilä, E. A.,** Lipoprotein lipase activity in adipose tissue and in postheparin plasma in human obesity, *Acta Med. Scand.*, 202, 399, 1977.
84. **Björntorp, P., Enzi, G., Ohlson, R., Persson, B., Sponbergs, P., and Smith, U.,** Lipoprotein lipase activity and uptake of exogenous triglycerides in fat cells of different size, *Horm. Metab. Res.*, 7, 230, 1975.
85. **Gruen, R., Hietanen, E., and Greenwood, M. R. C.,** Increased adipose tissue lipoprotein lipase activity during the development of the genetically obese rat (fa/fa), *Metabolism*, 27(Suppl. 2), 1955, 1978.

86. **Guy-Grand, B. and Bigorie, B.,** Effect of fat all size, restrictive diet and diabetes on lipoprotein lipase release by human adipose tissue, *Horm. Metab. Res.,* 7, 471, 1975.
87. **Pykälistö, O. J., Smith, P., and Brunzell, J. D.,** Determinants of human adipose tissue lipoprotein lipase: effects of diabetes and obesity on basal and diet induced activity, *J. Clin. Invest.,* 56, 1108, 1975.
88. **Taskinen, M.-R. and Nikkilä, E. A.,** Effects of caloric restriction on lipid metabolism in man. Changes of tissue lipoprotein lipase activities and of serum lipoproteins, *Atherosclerosis,* 32, 289, 1979.
89. **Lithell, H., Boberg, J., Hellsing, K., Lundqvist, G., and Vessby, B.,** Lipoprotein-lipase activity in human skeletal muscle and adipose tissue in the fasting and the fed states, *Atherosclerosis,* 30, 89, 1978.
90. **Keys, A., Grande, F., and Anderson, J. T.,** Fiber and pectin in the diet and serum cholesterol concentration in man, *Proc. Soc. Exp. Biol. Med.,* 106, 555, 1961.
91. **Kay, R. M. and Truswell, A. S.,** Effect of citrus pectin on blood lipids and fecal steroid excretion in man, *Am. J. Clin. Nutr.,* 30, 171, 1977.
92. **Mathur, K. S., Khan, M. A., and Sharma, R. D.,** Hypocholesterolaemic effect of bengal gram: a long-term study in man, *Br. Med. J.,* 1, 30, 1968.
93. **Flanagan, M., Little, C., Milliken, J., Wright, E., McGill, A. R, Weir, D. G., and O'Moore, R. R.,** The effects of diet on high density lipoprotein cholesterol, *J. Human Nutr.* 34, 43, 1980.
94. **Truswell, A. S.,** Food fibre and blood lipids, *Nutr. Rev.,* 35, 51, 1977.
95. **Zilversmit, D. B.,** Dietary fiber, in *Nutrition, Lipids, and Coronary Heart Disease,* Levy, R., Rifkind, B., Dennis, B., and Ernst, N., Eds., Raven Press, New York, 1979, 149.
96. **Bobek, P. and Ginter, E.,** Serum triglycerides and post-heparin lipolytic activity in guinea-pigs with latent vitamin C deficiency, *Experientia,* 34, 1554, 1978.
97. **Forfar, J. O. and Thompsett, L.,** Idiopathic hypercalcaemia of infancy, *Adv. Clin. Chem.,* 2, 167, 1959.
98. **Bierenbaum, M. L., Fleischman, A. I., and Reichelson, R. I.,** Long term human studies on the lipid effects of oral calcium, *Lipids,* 7, 202, 1972.
99. **Heyden, S., Heiss, G., Manegold, C., Tyroler, H. A., Hames, C. G., Bartel, A. G., and Cooper, G.,** The combined effect of smoking and coffee drinking on LDL and HDL cholesterol, *Circulation,* 60, 22, 1979.
100. **Studlar, M.,** Über det Einfluss von Coffein auf den Fettund Kohlenhydratstoffwechsel des Menschen, *Z. Ernährungswissen,* 12, 109, 1973.
101. **Glueck, C. J., Mattson, F., and Bierman, E. L.,** Diet and coronary heart disease: another view, *N. Engl. J. Med.,* 298, 1474, 1978.
102. **Scanu, A. M.,** Plasma lipoproteins and coronary heart disease, *Ann. Clin. Lab. Sci.,* 3, 79, 1978.

Chapter 5

HORMONES IN THE REGULATION OF LIPOPROTEIN METABOLISM

Eino Hietanen

TABLE OF CONTENTS

I. INTRODUCTION

Hormones are important factors in the regulation of the lipoprotein metabolism and lipoprotein levels in serum. They may modify the lipoprotein synthesis or their catabolism. The enhanced adipose tissue lipolysis produces increasing amounts of circulating fatty acids to be used in the liver for the lipoprotein synthesis. Simultaneous inhibition of the lipoprotein lipase enzyme may decrease the catabolism of existing lipoproteins, preventing their hydrolysis and accumulation in the adipose tissue or in the muscle tissue to be used as an energy source. The effect of individual hormones on serum lipoprotein fractions may vary markedly from one fraction to another (Table 1).

II. SEX HORMONES

Oral steroid hormones, estrogens and progestins, either in contraceptives or not, influence the serum high density lipoproteins (HDL) levels (Table 1). Female sex hormones, mainly estrogen, contribute to the elevated level of HDL cholesterol while male sex hormones decrease HDL levels.[1,2] The use of oral contraceptives may decrease HDL cholesterol levels.[1] Estrogen-containing contraceptives are known to elevate triglyceride levels in healthy women and to aggravate preexisting hypertriglyceridemia leading even to the development of pancreatitis.[3]

The net effect of combination pills depends on its formulation.[4] In a California study, women using oral contraceptives had higher plasma triglyceride levels and both systolic and diastolic blood pressures than those not using them when adjusted to age and leanness.[5] Similar increases, although less in magnitude, were present in those women using pure estrogens.

It has been suggested that the increased triglyceride levels caused by pills might be due to the increased synthesis of triglycerides.[6–8] Some experimental studies have suggested this to be due to the decreased lipoprotein lipase activity, i.e., due to the decreased triglyceride removal.[9] Hamosh and Hamosh[9] studied the effect of 17-β-estradiol and progesterone on the adipose tissues lipoprotein lipase activity in male and ovarectomized female rats. In the male rats 17-β-estradiol decreased the lipoprotein lipase activity from 8.22 to 4.96 U/g. Ovarectomy increased the activity from 10.4 to 22.7 U/g and the administration of estradiol (2.5 μg/day) decreased it back to 9 U/g and the dose of 25 μg/day to 3.2 U/g. In the ovarectomized rats serum triglycerides increased from 0.8 to 1.4 μmol/mℓ in 25 μg/day treated group. Progesterone did not affect the adipose tissue lipoprotein lipase activity in male or female rats. Estradiol treatment in experimental animals has opposite effects on the adipose tissue and cardiac lipoprotein lipases.[10] In the heart, the estradiol treatment increases the lipoprotein lipase activity while in the adipose tissue the activity decreases.[10]

In humans a decreased activity of the postheparin plasma lipolysis has been demonstrated both during the use of oral contraceptives and estrogen therapy.[7,11] Applebaum et al.[12] studied the effect of a 2-week ethinyl estradiol treatment on the extrahepatic lipoprotein lipase and hepatic triglyceride lipase in postheparin plasma. They demonstrated that the decrease due to estrogens in postheparin lipolytic activity was because of the decrease of the hepatic lipase activity in contrast to experimental studies.[9]

Glueck et al.[13] studied the effect of estrogens on the hepatic and extrahepatic lipases using protamin inactivation in hypertriglyceridemic subjects. In six women having conjugated estrogens the protamin inactivated triglyceride lipase (lipoprotein lipase) was decreased in a 2-weeks period; when estrogens were discontinued the protamin-inactivated lipoprotein lipase remained low while protamin resistant hepatic lipase increased. Mean triglycerides fell from 628 mg % to 447 mg % during the estrogen

Table 1
THE EFFECT OF HORMONES ON PLASMA LIPOPROTEINS

	HDL	LDL	VLDL	Chylomicrons	Total triglycerides	Total cholesterol
Estrogens	↑		↑		↑	
Progestins	↓				↓	
Oral contraceptives			↑		↑	
Glucocorticoids	⇀ ?	↓	↑		↑	↑ ⇀
Testosterone	↓					↑
Thyroxin					↓	
Insulin	↓ ⇀	↓	↓	↓	↓	↓

Note: ↑ = increase; ↓ = decrease; ⇀ no change; ? = no definite data.

treatment. There was no significant correlation between changes in the hepatic triglyceride lipase and triglycerides when estrogens were dicontinued. In another study a comparison of blood lipids was made in menopausal women between those using estrogens and those without drugs.[14] In the estrogen users the HDL_2 levels were higher than in nonusers. The difference was similar to that found between nonusers and age-matched males. Users of the combination pills had increased HDL_3 levels. Progestin effect was studied in hyperlipoproteinemic (type V) women (two) when noretindrone acetate decreased triglyceride levels, and at the same time there was a reduction in the HDL level, mainly in less dense species (HDL_2).[14] In this study HDL levels correlated in users and nonusers with total serum triglycerides. In men the group with high testosterone also had high HDL cholesterol and there was a positive correlation between HDL cholesterol and testosterone, and on the other hand between total cholesterol and testosterone.[15]

III. INSULIN

Insulin has both direct effects on plasma lipoproteins and lipoprotein metabolism and it may mediate the influence of other factors, e.g., diet and other hormones. The effect of insulin on plasma lipoproteins in man has often been studied in diabetes. In diabetics serum triglyceride and very low density lipoproteins (VLDL) levels are higher than normal.[16] Treatment with insulin decreases VLDL levels and total triglycerides.[16,17] Insulin mediates its effects probably either through the synthesis of lipoproteins or by the regulation of their catabolism.

The lipoprotein lipase (LPL) enzyme is active in the capillary endothelium and hydrolyses VLDL and chylomicron triglycerides. Possibly this extracellular form originates from intracellular LPL. The LPL enzyme has a very short half-life and is thus dependent on the continuous protein synthesis.[18] Even a short-term fasting can markedly decrease the LPL activity but when refeeding after fasting the LPL activity correlates well with the plasma insulin levels.[18] In vitro insulin has been found to increase the LPL activity in starved rats.[19] Insulin might possibly enhance intracellular-extracellular interconversion of LPL and consequently the LPL secretion.[4,20–22]

Garfinkel et al.[21] studied the regulation of LPL activity by insulin using fasting rats. Fasting decreases the LPL activity very rapidly but when insulin was administered in vivo the epididymal fat pad LPL activity increased rapidly. Results suggested that the increased adipose tissue LPL activity results mainly from the enhanced enzyme secretion by fat cells proposing that the secretory mechanisms are the major sites for the LPL activity regulation in adipocytes. In studies with diabetic insulin treated patients Nikkilä and co-workers[16] demonstrated rather convincingly that LPL might be the key factor in the regulation of serum lipoproteins.[16,23] Insulin treatment increased the enzyme activity and facilitated lipoprotein catabolism. Although hormone sensitive lipase may not be as important in the regulation of serum lipoproteins as the LPL enzyme, it might also contribute, to some extent, this regulation.

Olefsky[24] compared the ability of insulin to inhibit the epinephrine-stimulated lipolysis in large and small rat adipocytes. The large cells were obtained from older, obese animals, and small ones were from younger, lean animals. Large adipocytes were found to be less sensitive to antilipolytic effects of insulin than small ones. The dose-response curve to insulin was shifted to the right for insulin's antilipolytic action and is consistent with the theory on decreased number of insulin receptors in older ages. This might, in part, contribute to the development of the maturity-onset diabetes where LPL activity has less role.[16]

Thus, to conclude, in addition to in vitro and experimental studies, diabetes has also proved valuable in the evaluation of the insulin effects on lipid metabolism. It is typical that especially obese diabetics have high serum total triglyceride levels[25] and low triglyceride clearance. The administration of insulin improves the situation drastically by decreasing serum triglyceride levels and by increasing lipolytic activity.[25] Moreover, at least experimentally, insulin increases muscle triglyceride utilization for energy needs.[26]

IV. GLUCOCORTICOIDS

Glucocorticoids are known to decrease LDL fraction of lipoproteins while total triglyceride concentration and VLDL concentration are elevated[27,28] (Table 1). They also possibly mediate their influence on serum lipoproteins by the LPL enzyme (Table 2). Bagdade et al.[27] studied the effect of the glucocorticoid administration on serum lipids and the mechanisms of action. They injected dexamethasone (0.125 mg/kg) daily i.m. for 2 weeks to male Spraque-Dawley rats. Significant increases in total triglycerides and VLDL triglycerides were found without any changes in the cholesterol concentration and the actual decrease in LDL triglycerides and cholesterol were found. The dexamethasone treatment was associated with the increased basal insulin and glucose levels and a reduction in the LPL activity in epididymal fat pads. The results suggested that the hypertriglyceridemia is primarily due to the decreased removal of VLDL triglycerides because of the lowered LPL activity. This result is quite controversial considering that increased insulin levels are related to the enhanced LPL activity and the result is also conflicting with that of de Gasquet et al.[29] They found that the adipose tissue LPL activity was increased both in fed and fasted rats by glucocorticoid injection and decreased by adrenalectomy. The heart LPL activity was increased by a single injection of glucocorticoid to the same high level as found in fasting rats. The rise of the activity by fasting in the heart was reduced by adrenalectomy. Thus, the nutritional effects on LPL activity may also partly be mediated by adrenal hormones.

The glucocorticoid administration is connected to the changes in other hormones. Increased levels of plasma immunoreactive insulin, glucose, and free fatty acids were present both at fasting and after i.v. glucose load after glucocorticoid administration, while no changes in the postheparin serum lipolytic activity were present in this study.[28] These results indicate that glucocorticoid-induced hypertriglyceridemia is primarily due to the increased hepatic triglyceride production rate.[28] It is quite possible that the acute

Table 2
THE HORMONAL REGULATION OF THE LIPOPROTEIN LIPASE ENZYME ACTIVITY

Hormone	Postheparin plasma lipolytic activity and adipose tissue	Muscle tissue
Glucocorticoids	↓ ? ↑	↑
Insulin	↑	↑
Epinephrine	↓	
Progesterone	⇌	⇌ ↑
Estrogens	↓	
Oral contraceptives	⇌	
Thyroxin	↑	
ACTH	↓	
Diabetes	↓	↓ ↑ ⇌ ?
Hypothyroidism	↓	
Adrenalectomy	↓	↓
Ovarectomy	↑	

Note: ↑ = increase; ↓ = decrease; ⇌ = no change; ? = no definite data.

effects of glucocorticoids on serum lipoproteins are mediated through different mechanisms than the long-term effects which might partly explain different results.

V. OTHER HORMONES

It is well-known that hypothyroidism is characterized, e.g., with increased plasma triglyceride concentration and that treatment with thyroxin hormone decreases triglyceride levels although the exact way of influence has been biochemically less well defined. Pykälistö et al.[30] studied the influence of hypothyroidism on the lipoprotein lipase activity and its role in the regulation of triglyceride metabolism in six hypothyroid patients with and without L-thyroxin treatment. In hypothyroid patients the heparin eluable LPL was lower in adipose tissue biopsies than in controls but increased with treatment 163%. The total activity, when measured in ammonium hydroxide extracts of acetone-ether powders, was not statistically different from the controls and did not increase with treatment. Also the plasma postheparin lipolytic activity was low in the hypothyroid patients and increased with treatment (111%); this increase correlated well with the increase in the activity of the heparin eluable LPL. Plasma triglyceride levels decreased during the treatment 43% and the changes correlated negatively with the changes in the heparin eluable LPL activity suggesting that thyroxin hormone mediates its effects via this enzyme.

Although the effects of catecholamines and ACTH on plasma lipids are not well characterized, they might also modulate these. Serum ACTH may have its effects indirectly mediated by the changes in glucocorticoid secretion. However, both enzymes have also been found to influence the LPL (Table 2) and hormone-sensitive lipase activities (Table 3). They have been found to have opposite effects to insulin on the intracellular LPL.[31]

Hormone sensitive lipolytic activity plays a minor role in the metabolism of lipoproteins but it may contribute indirectly to lipoprotein metabolism, either producing free fatty acids from adipose tissue triglycerides or by regulating the overall triglyceride

Table 3
THE REGULATION OF THE HORMONE SENSITIVE LIPASE IN ADIPOSE TISSUE

Increased activity	Decreased activity
Catecholamines	Insulin
Glucagon	Prostaglandins
ACTH	
Growth hormone	
Glucocorticoids	
Thyroid hormone	

turnover in adipose tissue. The activity of the hormone-sensitive lipase is regulated by the same hormones as that of the LPL but to the opposite direction (Table 3). Typical to this enzyme is the stimulation of the activity by catecholamines.[32,33] Catecholamines probably mediate their effect by the cell surface receptor which then stimulates the cyclic AMP-dependent protein kinase pathway.[34–37] Both ACTH and glucagon are potent activators of hormone-sensitive lipase, especially at young ages.[4,38–41] In adipocytes insulin decreases intracellular norepinephrine stimulated lipolysis probably by the inhibition of hormone sensitive lipase.[42,43]

In addition to numerous hormones stimulating hormone-sensitive lipase,[44] drugs may interfere with this activity. At least in vitro propranolone inhibits the norepinephrine-stimulated lipolysis in adipocytes.[44] Also in vivo in man β-blocking agents probably inhibit adipose tissue lipolysis as judged from decreased serum free fatty acid levels due to alprenolol treatment.[45,46] The mobilization of free fatty acids from triglycerides in adipose tissue by the hormone-sensitive lipase may also yield increased production of VLDL lipoproteins in the liver and thus even the hormone-sensitive lipase may contribute to the regulation of the lipoprotein metabolism.[47] The apoprotein moiety of high density lipoproteins interacts with the hormone-sensitive lipase complex suggesting indirectly that this enzyme might also be involved in the lipoprotein metabolism.[48–51]

REFERENCES

1. **Arntzenius, A. C., van Gent, C. M., van der Voort, H., Stegerhoek, C. I., and Styblo, K.,** Reduced high-density lipoprotein in woman aged 40–41 using oral contraceptives, *Lancet,* 2, 1221, 1978.
2. **Nikkilä, E. A.,** Metabolic and endocrine control of plasma high density lipoprotein concentration. Relation to catabolism of triglyceride-rich lipoproteins, in *High Density Lipoproteins and Atherosclerosis,* Gotto, A. M., Jr., Miller, N. E., and Oliver, M. F., Eds. Elsevier/North Holland Biomedial Press, Amsterdam, 1978, 177.
3. **Glueck, C. J., Scheel, D., Fishback, J., and Steiner, P.,** Estrogen induced pancreatitis in patients with previously covert familial type V hyperlipoproteinemia, *Metab. Clin. Exp.,* 21, 657, 1972.
4. **Saggerson, E. D., Sooranna, S. R., and Harper, R. D.,** Hormones and fatty acid metabolism in white adipose tissue, *Biochem. Soc. Trans.,* 5, 900, 1977.
5. **Stern, M. P., Brown, B. W., Jr., Haskell, W. L., Farquhar, J. W., Wehrle, C. L., and Wood, P. D. S.,** Cardiovascular risk and use of estrogens or estrogen-progestagen combinations. Stanford Three-Community Study, *JAMA,* 235, 811, 1976.
6. **Glueck, C. J., Fallet, R. W., and Scheel, D.,** Effects of estrogenic compounds on triglyceride kinetics, *Metab. Clin. Exp.,* 24, 537, 1975.

7. **Hazzard, W. R., Brunzell, J. D., Notter, D. T., Spiger, M. J., and Bierman, E. L.,** Estrogens and triglyceride transport: increased endogenous production as the mechanism for the hypertriglyceridemia of oral contraceptive therapy, *Endocrinol. Excerpta Med. Int. Congr. Ser.*, 273, 1006, 1973.
8. **Kekki, M. and Nikkilä, E. A.,** Plasma triglyceride turnover during use of oral contraceptives, *Metab. Clin. Invest.*, 20, 878, 1971.
9. **Hamosh, M. and Hamosh, P.,** The effect of estrogen on the lipoprotein lipase activity of rat adipose tissue, *J. Clin. Invest.*, 55, 1132, 1975.
10. **Wilson, D. E., Flowers, C. M., Carlile, S. I., and Udall, K. S.,** Estrogen treatment and gonadal function in the regulation of lipoprotein lipase, *Atherosclerosis*, 24, 491, 1976.
11. **Hazzard, W. R., Spiger, M. J., Bagdade, J. D., and Bierman, E. L.,** Studies on the mechanism of increased plasma triglyceride levels induced by oral contraceptives, *N. Engl. J. Med.*, 280, 471, 1969.
12. **Applebaum, D. M., Goldberg, A. P., Pykälistö, O., Brunzell, J. D., and Hazzard, W. R.,** Effect of estrogen on postheparin lipolytic activity. Selective decline in hepatic triglyceride lipase, *J. Clin. Invest.*, 59, 601, 1977.
13. **Glueck, C. J., Gartside, P., Fallat, R. W., and Mendoza, S.,** Effect of sex hormones on protamine inactivated and resistant postheparin plasma lipases, *Metabolism*, 25, 625, 1976.
14. **Krauss, R. M., Lindegren, F. T., Wingerd, J., Bradley, D. D., and Ramcharan, S.,** Effects of estrogens and progestins on high density lipoproteins, *Lipids*, 14, 113, 1979.
15. **Nordøy, A., Aakvaag, A., and Thelle, D.,** Sex hormones and high density lipoproteins in the healthy males, *Atherosclerosis*, 34, 431, 1979.
16. **Nikkilä, E. A., Huttunen, J. K., and Ehnholm, C.,** Postheparin plasma lipoprotein lipase and hepatic lipase in diabetes mellitus, *Diabetes*, 26, 11, 1977.
17. **Taskinen, M-R. and Nikkilä, E. A.,** Lipoprotein lipase activity of adipose tissue and skeletal muscle in insulin-deficient human diabetes, *Diabetologia*, 17, 351, 1979.
18. **Cryer, A., Riley, S. E., Williams, E. R., and Robinson, D. S.,** Effects of fructose, sucrose and glucose feeding on plasma insulin concentrations and on adipose tissue clearing factor lipase activity in the rat, *Biochem. J.*, 140, 561, 1974.
19. **Robinson, D. S. and Wing, D. R.,** in *Adipose Tissue*, Jeanrenaud, B. and Hepp, D., Eds., Academic Press, New York, 1970, 41.
20. **Cryer, A., Riley, S. E., William, E. R., and Robinson, D. S.,** Effect of nutritional status on rat adipose tissue, muscle and posthepatin plasma clearing factor lipase activities: their relationship to triglyceride fatty acid uptake by fat-cells and to plasma insulin concentrations, *Clin. Sci. Mol. Med.*, 50, 213, 1976.
21. **Garfinkel, A. S., Nilsson-Ehle, P. and Schotz, M. C.,** Regulation of lipoprotein lipase induction by insulin, *Biochim. Biophys. Acta*, 424, 264, 1976.
22. **Nilsson-Ehle, P., Garfinkel, A. S., and Schotz, M. C.,** Intracellular and extracellular forms of lipoprotein lipase in adipose tissue, *Biochim. Biophys. Acta*, 431, 147, 1976.
23. **Pykälistö, O., Smith, P. H., and Brunzell, J. D.,** Determinants of human adipose tissue lipoprotein lipase. Effect of diabetes and obesity on basal- and diet-induced activity, *J. Clin. Invest.*, 56, 1108, 1975.
24. **Olefsky, J. M.,** Insensitivity of large rat adipocytes to the antilipolytic effects of insulin, *J. Lipid Res.*, 18, 459, 1977.
25. **Taylor, K. G., Galton, D. J., and Holdsworth, G.,** Insulin-independent diabetes: a defect in the activity of lipoprotein lipase in adipose tissue, *Diabetologia*, 16, 313, 1979.
26. **Stearns, S. B., Tepperman, H. M., and Tepperman, J.,** Studies on the utilization and mobilization of lipid in skeletal muscles from streptozotocin-diabetic and control rats, *J. Lipid Res.*, 20, 654, 1979.
27. **Bagdade, J. D., Yee, E., Albers, J., and Pykälistö, O.,** Glucocorticoids and triglyceride transport: effects on triglyceride secretion rates, lipoprotein lipase, and plasma lipoproteins in the rat, *Metabolism*, 25, 533, 1976.
28. **Mibayashi, Y.,** Glucocorticoid-induced hypertriglyceridemia: effects of cortisone acetate on triglyceride secretion rates and post-heparin lipolytic activity in rabbits, *Folia Endocrinol. Jpn.*, 54, 203, 1978.
29. **de Gasquet, P., Pequignot-Planche, E., Tonnu, N. T., and Diaby, F. A.,** Effect of glucocorticoids on lipoprotein lipase activity in rat heart and adipose tissue, *Horm. Metab. Res.*, 7, 152, 1975.
30. **Pykälistö, O., Goldberg, A. P., and Brunzell, J. D.,** Reversal of decreased human adipose tissue lipoprotein lipase and hypertriglyceridemia after treatment of hypothyroidism, *J. Clin. Endocrinol. Metab.*, 43, 591, 1976.
31. **Davis, P., Cryer, A., and Robinson, D. S.,** Hormonal control of adipose tissue clearing factor lipase, *FEBS Lett.*, 45, 271, 1974.

32. **Hartman, A. D., Cohen, A. I., Richane, C. J., and Hsu, T.,** Lipolytic response and adenyl cyclase activity of rat adipocytes as related to cell size, *J. Lipid Res.*, 12, 498, 1971.
33. **Khoo, J. C., Severson, D. L., and Steinberg, D.,** Reversible deactivation of chicken adipose tissue hormone-sensitive lipase by rat liver phosphorylase phosphatase, *Fed. Proc. Fed. Am. Soc. Exp. Biol.*, 35, 1412, 1976.
34. **Belfrage, P., Jergil, B., Strålfors, P., and Tornqvist, H.,** Hormone-sensitive lipase of rat adipose tissue: identification and some properties of the enzyme protein, *FEBS Lett.*, 75, 259, 1977.
35. **Huttunen, J. K. and Steinberg, D.,** Activation and phosphorylation of purified adipose tissue hormone-sensitive lipase by cyclic AMP-dependent protein kinase, *Biochim. Biophys. Acta,* 239, 411, 1971.
36. **Khoo, J. C., Aquino, A. A., and Steinberg, D.,** The mechanism of activation of hormone-sensitive lipase in human adipose tissue, *J. Clin. Invest.*, 53, 1124, 1974.
37. **Khoo, J. C. and Steinberg, D.,** Reversible protein kinase activation of hormone-sensitive lipase from chicken adipose tissue, *J. Lipid Res.*, 15, 602, 1974.
38. **Cooper, B. and Gregerman, R. I.,** Hormone-sensitive fat cell adenylate cyclase in the rat. Influences of growth, cell size, and aging, *J. Clin. Invest.*, 57, 161, 1976.
39. **Cooper, B., Weinblatt, F., and Gregerman, R. I.,** Enhanced activity of hormone-sensitive adenylate cyclase during dietary restriction in the rat. Dependence on age and relation to cell size, *J. Clin. Invest.*, 59, 467, 1977.
40. **Grundleger, M. L., and Bernstein, R. S.,** Augmental lipolysis in rat adipose tissue upon repeated exposure to epinephrine, *Metabolism,* 28, 989, 1979.
41. **Jackson, R. L., Chan, L., Snow, L. D., and Means, A. R.,** Hormonal regulation of lipoprotein synthesis, in *Disturbances in Lipid and Lipoprotein Metabolism,* Dietschy, J. M., Gotto, A. M., Jr., and Ontko, J. A., Eds., American Physiological Society, Bethesda, Md., 1978, 139.
42. **Burns, T. W., Terry, B. E., Langley, P. E., and Robinson, G. A.,** Insulin inhibition of lipolysis of human adipocytes. The role of cyclic adenosine monophosphate, *Diabetes,* 28, 957, 1979.
43. **Garnie, J. A., Smith, D. G., and Mayris-Vavayannis, M.,** Effects of insulin on lipolysis and lipogenesis in adipocytes from genetically obese (ob/ob) mice, *Biochem. J.*, 184, 107, 1979.
44. **Steinberg, D. and Khoo, J. C.,** Hormone-sensitive lipase of adipose tissue, *Fed. Proc. Fed. Am. Soc. Exp. Biol.*, 36, 1986, 1977.
45. **Frisk-Holmberg, M., Jorfeldt, L., and Juhlin-Dannfeldt, A.,** Influence of alprenolol on hemodynamic and metabolic responses to prolonged exercise in subjects with hypertension, *Clin. Pharmacol. Ther.*, 21, 675, 1977.
46. **Nilsson, A., Hansson, B.-G., and Hökfelt, B.,** Effect of metoprolol on blood glycerol, free acids, triglycerides and glucose in relation to plasma catecholamines in hypertensive patients following submaximal work, *Eur. J. Clin. Pharmacol.*, 13, 5, 1978.
47. **Steinberg, D.,** Hormonal control of lipolysis in adipose tissue, *Adv. Exp. Med. Biol.*, 26, 77, 1972.
48. **Fain, J. N.,** Biochemical aspects of drug and hormone action on adipose tissue, *Pharmacol. Rev.*, 25, 67, 1973.
49. **Fain, J. N.,** Inhibition of adenosine cyclic 3′,5′-monophosphate accumulation in fat cells by adenosine, N^6 (phenylisopropyl) adenosine, and related compounds, *Mol. Pharmacol.*, 9, 595, 1973.
50. **Fain, J. N., Shepherd, R. E., Malbon, C. C., and Moreno, E. J.,** Hormonal regulation of triglyceride breakdown in adipocytes, in *Disturbances in Lipid and Lipoprotein Metabolism,* Dietschy, J. M., Gotto, A. M., Jr., and Ontko, J. A., Eds., American Physiological Society, 1978, 213.
51. **Khoo, J. C., Drevon, C. A., and Steinberg, D.,** Dissociation of the lipid-enzyme complex of hormone-sensitive lipase using high density lipoprotein or apolipoprotein A-I, *Biochim. Biophys, Acta,* 617, 540, 1980.

Chapter 6

ENVIRONMENTAL FACTORS AND LIPOPROTEINS

Eino Hietanen

TABLE OF CONTENTS

I. DRUGS

Hyperlipoproteinemias are important genetic or secondary disorders of lipoprotein metabolism which may be a problem prognostically in terms of the development of coronary heart disease.[1–3] The development of drugs lowering blood lipids has led to the treatment of the hyperlipoproteinemic states. In addition to the drugs aimed at lowering blood lipid levels, some drugs may cause secondary changes in plasma lipoproteins although they have been aimed for other diseases. It is of importance to know the effects of drugs on serum lipoprotein patterns as well as the mechanisms of how these possible changes are mediated by these drugs. It might be of importance prognostically that such drugs meant to lower possible risk factors for coronary heart disease, e.g., β-blocking agents, do not, on the other hand, decrease HDL cholesterol, and this decrease might increase the risk for coronary heart disease.

A. Antihyperlipidemic Drugs

In addition to cholesterol, high triglyceride levels in serum also may increase the risk for the coronary heart disease and atherosclerosis.[4] The role of hypertriglyceridemia as an independent risk factor has not been conclusively proved.[5,6] Even if the high triglyceride level very low density lipoprotein (VLDL, fraction of lipoproteins) might not be an independent risk factor, the inverse relationship of high plasma triglyceride concentration and high density lipoprotein (HDL) concentration supports the hypothesis that it is favorable to try to decrease triglyceride concentration.

Hyperlipoproteinemias are usually classified into five types according to the changes in various lipoprotein fractions which is also reflected in the amount of cholesterol or triglyceride moieties in serum (Table 1).[7] A summary of these types is given in Table 1. Most common hyperlipoproteinemias are types II and IV. Type II hyperlipoproteinemia is characterized by the increased serum cholesterol and triglyceride levels and high concentration of cholesterol in various tissues, including the coronary arteries.[8] The metabolic disorder is probably the absence of low density lipoprotein (LDL) receptors in the peripheral cells.[9,10] Nikkilä et al.[11] found a low activity of plasma postheparin hepatic lipase activity in patients with familial type II hypercholesterolemia while the lipoprotein lipase activity was normal. Type I hyperlipoproteinemia is characterized by the decreased activity of the lipoprotein lipase enzyme which leads to the decreased removal of chylomicron triglycerides.[8]

As the purpose of this chapter is not to deal in-depth with the treatment of the pathological disorders of lipid methabolism, a summary table (Table 2) is shown to give the possibilities in the drug management of hyperlipoproteinemias. The mechanisms of how lipid-lowering drugs decrease plasma lipoprotein cholesterol and triglycerides have not been totally clarified.[12] One of the most widely used drugs to treat hyperlipoproteinemias is clofibrate (*p*-chlorophenoxisobutyrate) which reduces elevated levels of total cholesterol and triglycerides.[13–16] Clofibrate is known to lower serum triglyceride and cholesterol levels, possibly by increasing the peripheral removal of triglycerides.[17–19] One possible factor increasing the peripheral removal of triglycerides is the activation of the adipose tissue lipoprotein lipase activity. In the study by Tolman et al.[18] they found in experimental animals the activation of the adipose tissue lipoprotein lipase. Also, in other studies, similar results have been found both in animals and in man.[20–22] Greten et al.[23] found that clofibrate specifically increased the lipoprotein lipase activity and not the hepatic lipase activity. However, e.g., the binding of lipid-lowering drugs to the lipoprotein molecules may alter their structure and modify their catabolism.[24] Yet other possibilities exist such as decreased absorption of lipids, reduced production of LDL by the cells and lowered secretion of lipoproteins.

Table 1
HYPERLIPOPROTEINEMIA CLASSIFICATION

Type	Lipoproteins fasting chylomicrons	LDL	VLDL	HDL	Lipids cholesterol	Triglycerides	Primary cause	Other major causative diseases
I	Present	↔ ↓	↔, ↑, ↓	⇊	↔, ↑	⇈	Familial lipoprotein lipase deficiency	Insulin deficient diabetes, hypothyroidism
IIA	Not present	↑	↔, ↓	↔	↑	↔	Familial hypercholesterolemia	Hypothyroidism, kidney failure Dietary factors, liver diseases
IIB	Not present	↑	↑	↔	↑	↑	Familial hypercholesterolemia	Dietary factors, liver diseases
III	Present	IDL ↑, LDL_2 ↓	↑	↔, ↓	↑	↑	Familial dyslipoproteinemia	Rare
IV	Absent	↔, ↓	↑	↔, ↓	↔, ↑	↑	Familial hypertriglyceridemia	Diabetes, hypothyroidism, liver diseases, kidney failure
V	Present	↔, ↓	↑	↓	↔, ↑	↑ ↑	Decreased lipoprotein lipase activity	Drugs

Note: Symbols: ↔ = normal, ↓ = decreased, ⇊ = highly decreased, ↑ = increased, ⇈ = greatly increased.

Table 2
THERAPEUTIC DRUGS AND HYPERLIPOPROTEINEMIAS

	Clofibrate	Cholestipol	Ciprofibrate	Bezafibrate	Probucol	Nicotinic Acid	Cholestyram
LDL	↓	↓	⇊	↓	↓	↓	⇊
VLDL	↓	⇀	⇊	↓		↓	
HDL	↑	↑	↔	↑	↓	↑	⇁
Total triglycerides	↓	↑	↓	↓	↔	⇊	⇈
Total cholesterol	↓	↓	↓	↓	↓	↓	↓

Note: Symbols: ↔ no effect; ↓ decrease; ⇊ marked decrease; ↑ increase; ⇈ great increase.

In the recent study by Goldberg et al.[25] on patients having hypertriglyceridemia due to chronic renal failure and treated with hemodialysis, the effects of clofibrate on plasma lipids, and adipose and postheparin plasma lipoprotein lipase activities were studied. They found that clofibrate decreased total serum triglycerides as well as VLDL triglycerides and cholesterol and increased HDL cholesterol. This study demonstrated rather convincingly that clofibrate increases the peripheral removal of triglycerides which is at the same time connected with the increased formation of HDL.[26–29] Goldberg et al.[25] found a significant increase of the adipose tissue and postheparin serum lipoprotein lipase activities while no changes in the hepatic lipase activity due to clofibrate. The changes of the lipoprotein lipase activities and those of serum lipoproteins correlated well with each other suggesting a firm link between these parameters. It is possible that a decreased lipoprotein lipase activity may be one of the factors leading to the appearance of hyperlipoproteinemias. Although vitamin E may have many important functions in the body related to its properties as a powerful antioxidant, it does not have any marked additional effects on diet and lipid-lowering drugs in the regulation of plasma lipoproteins. In patients with hyperlipoproteinemia, Vessby et al.[30] found only a transient small change in the relative amounts of lipoprotein fatty acids due to the addition of vitamin E in the therapy.[30,31]

There is a continuous need for the development of new methods and drugs to treat lipid disorders. Many of the new drugs are related to clofibrate such as ciprofibrate and bezafibrate.[32] Ciprofibrate is useful in the treatment of type II and IV hyperlipoproteinemias by decreasing both total cholesterol and triglycerides.[31] This is mainly due to the decrease in VLDL triglycerides and LDL cholesterol while no decrease in HDL fraction is present. Ciprofibrate inhibits cholesterol biosynthesis in the liver. It may also have an effect by inhibiting the hormone sensitive lipase,[33,34] which suggests indirectly that it might activate the adipose tissue lipoprotein lipase (LPL) as well. It is also possible that hypolipemic drugs cause hydrolysis of cholesterol esters in the arterial walls, reducing the cholesterol accumulation and the progress of atherosclerosis.[12] The effects of bezafibrate are similar to clofibrate but this analogue is more potent than clofibrate and may cause the elevation of HDL cholesterol fraction in addition to the decreased LDL cholesterol and VLDL triglyceride levels.[32]

Recently attention has also been focused on drugs possibly altering the apoprotein moieties in lipoproteins.[35] Both in humans and in experimental animals apoprotein changes have been described following metformin treatment.[35] Experimentally metformin decreases atheromatous lesions in rabbits without much influence on the serum cholesterol levels.[36] This compound decreases apoprotein E in type III hyperlipoproteinemias in man and increases in HDL fraction apoA1/A2 ratio.[37] Metformin increases HDL cholesterol both in hyperlipoproteinemic patients and in normolipidemic persons.[35] Tiadenol has been studied in patients with type IV hyperlipoproteinemia where it decreases VLDL triglycerides and prevents carbohydrate induced increase of VLDL.[35] It also modifies the composition and structure of VLDL particles.

Para-aminosalisylic acid has also proved to be an effective drug in lowering serum lipids in types II and IV hyperlipoproteinemias[38,39] This drug decreases both total cholesterol and triglyceride levels significantly, fractions involved are VLDL and LDL lipoproteins[39] although in another study no change in LDL cholesterol was present.[38] Para-aminosalisylic acid apparently does not mediate its effects through changes in triglyceride catabolism catalyzed by the LPL enzyme but preferably through changes in triglyceride synthesis.[38,39]

Cholestipol is efficient in decreasing total plasma cholesterol and LDL cholesterol levels without any effects on HDL cholesterol while increasing total triglyceride levels.[40] Propucol did not decrease the total triglyceride concentration but was effective in decreasing total and LDL cholesterol levels, but it also decreased HDL cholesterol concentration by 22.6%[40] Both cholestipol and cholestyramine are effective in decreasing total plasma and LDL cholesterol levels in patients with type II a hyperlipoprotein emia.[40,41] However, this treatment increases total triglycerides if no combination therapy is given. No changes in HDL cholesterol levels were found. The trials on other resins are also rather common in attempts to decrease bile acid enterohepatic circulation and thus in increasing the consumption of cholesterol for the bile acid synthesis in the liver.[42] Nicotinic acid affects reducing cholesterol and triglyceride levels, mainly by inhibiting the synthesis of apolipoproteins in the liver and by reduction of their turnover, and has no effects on bile acid synthesis.[43,44] The combination of nicotinic acid in the hyperlipoproteinemia therapy might be beneficial even in the way that this compound, by decreasing the fractional catabolism of HDL, leads to the increased HDL cholesterol levels.[45]

B. Other Drugs

Apart from drugs used for the treatment of hyperlipoproteinemias, other drugs also intervene with plasma lipoproteins. Some of these effects have been dealt with in the section on the influence of hormones on serum lipoproteins, as hormones are also used as drugs to supplement endogenous metabolic disorders or for other reasons. These include such compounds as glucocorticoids, insulin, sex steroids, and thyroxin. In this section only nonhormonal drugs will be dealt with.

The main purpose is to deal with drugs possibly disordering the normal distribution of serum lipoprotein patterns. In this respect β-blocking drugs are among the first ones to be dealt with due to their large use in persons possibly having other risk factors than high cholesterol levels. These factors include hypertension which is often treated with β-blocking agents. These drugs are also used to prevent arrhythmias after myocardial infarction and in coronary heart disease. Both in experimental studies and in humans when β-blocking agents have been given, the inhibition of the adipose tissue lipolysis has been found which might interact with the exercise induced changes in the energy metabolism.[46–49] Due to the inhibition of lipolysis both in the adipose tissue and muscles, β-blocking agents may prevent or decrease the use of fatty acids as energy source.[46]

Recently, the effect of β-blockers on serum lipids in patients treated for hypertension or coronary heart disease has been studied.[50–51] β-Blockers decreased serum HDL cholesterol concentration while no changes in other lipoproteins were present. In another cross-sectional study no changes in serum HDL cholesterol were present in patients

Table 3
THE EFFECT OF A SINGLE DOSE OF β-BLOCKING DRUG ON PLASMA LIPIDS AND ADIPOSE TISSUE LIPOPROTEIN LIPASE ACTIVITY

	Before	After
Total cholesterol (mmol/ℓ)	5.18 ± 0.25	5.15 ± 0.29
HDL cholesterol (mmol/ℓ)	1.48 ± 0.08	1.47 ± 0.06
HDL/total ratio (%)	20.1 ± 1.7	29.4 ± 1.8
Total triglyceride (mmol/ℓ)	1.20 ± 0.09	1.17 ± 0.11
Lipoprotein lipase (μmol/h/g)	2.19 ± 0.27	1.87 ± 0.20

Note: One 160 mg tablet of sotalol was given after the initial fasting samples and 24 hr after the drug another fasting sample was taken from 11 young (22 to 34 years) male volunteers.

treated with β-blockers combined with diuretics or hydralazine.[52] In a longitudinal study by Pagnan et al.[53] the effect of labetalol on serum lipids was studied in patients treated for hypertension for 4 months by giving the drug orally up to the dose of 1200 mg daily. They did not find any significant changes in the total cholesterol or triglyceride concentrations but in the VLDL fraction the cholesterol/triglyceride ratio decreased, suggesting the relative increase of triglyceride rich particles.

We administered β-blocking agent, sotalol, to voluntary persons at a dose of 160 mg as a single dose and followed the effect of this dose on serum total triglycerides, cholesterol, HDL cholesterol, and the adipose tissue LPL activity (Table 3). In this study no changes in any of the serum lipid parameters were found. Neither did the change in the LPL activity correlate significantly with changes in serum lipoproteins. However, there was a significant negative correlation between the change in HDL cholesterol and HDL cholesterol (r = −0.58), and on the other hand, between the change in the adipose tissue lipoprotein lipase activity and lipoprotein lipase activity (r = −0.74) at the beginning of the experiment. In further experiments, we studied the effects of sotalol on serum lipids and adipose tissue lipoprotein lipase in patients treated for hypertension at least 2 years at a daily dose of 160 to 480 mg.[54] These patients interrupted the drug therapy under control for 3 weeks to evaluate the effect of the treatment on hypertension. No other medication was given. At the beginning and after a 3 week interval without sotalol, blood lipids and the adipose tissue LPL activity were determined. In these patients sotalol did not affect the total serum cholesterol concentration but there was a significant decrease in the triglycerides after the drug break and increase in the HDL cholesterol; however, no changes were present when the adipose tissue LPL activity was determined (Figure 1).

Interesting results were found when changes in the adipose tissue LPL activity were correlated with the changes in serum lipids.[54] During the drug treatment no correlation was found between the HDL cholesterol level and the adipose tissue LPL activity (r = −0.34, not significant by *t*-test). However, after 3 weeks break, there was a significant correlation between the HDL cholesterol and the adipose tissue LPL activity (r = +0.77, $p < 0.05$). Also, when serum triglycerides from patients without hypertriglyceridemia were correlated with the adipose tissue LPL activity a difference in this correlation was found during (r = 0.23) and after the break (r = −0.96) in drug dosage. These data suggest that β-blocking agents may intervene in the lipid metabolism by interacting with the LPL enzyme and thus by altering the HDL synthesis rate and consequently also triglyceride catabolism. The decreased HDL cholesterol levels during the treatment should be taken into account when administering β-blocking agents to

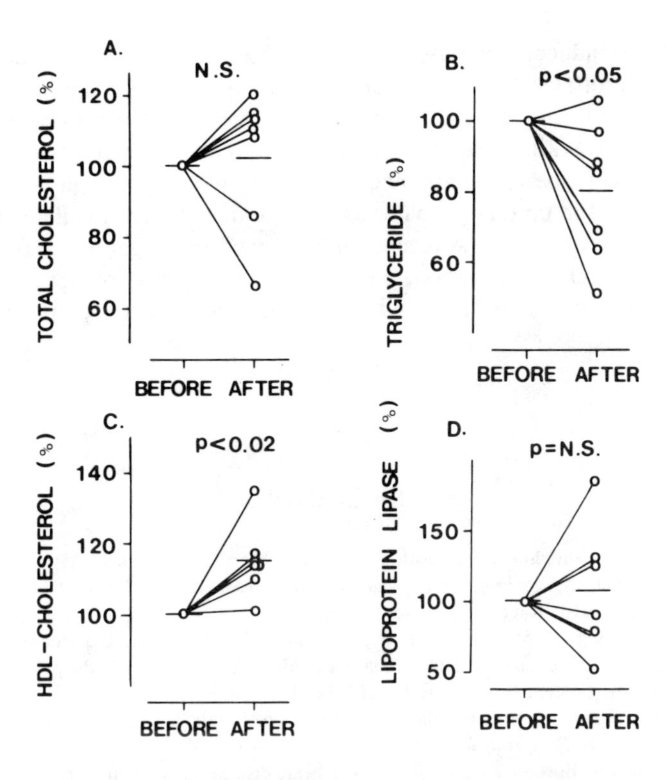

FIGURE 1. Serum total cholesterol (A), triglyceride (B), HDL cholesterol (C) concentrations and adipose tissue lipoprotein lipase activity (D) in seven hypertensive sotalol-using subjects before and after 3-week break in the β-blocking treatment. The significance of the differences between before and after levels and relative changes after the break in treatment are indicated in the figure.

persons with modest hypertension as these drugs might in the long run decrease the HDL cholesterol concentration which might increase the amount of risk factors for atherosclerosis.

In conclusion, the effects of β-blocking drugs on plasma lipoproteins need further studies. Possibly on the basis of these preliminary studies strict consideration should be used before prescribing these drugs in order not to hamper the advantageous effects by increasing the amount of total risk factors.

II. ENVIRONMENTAL COMPOUNDS

In addition to therapeutic drugs, lipoproteins are influenced by environmental contaminants as well. These may also intervene with the incidence of coronary heart disease and other atherosclerotic diseases although no definite epidemiological or experimental studies exist. Juchau et al.[55] have proposed that the mutations of smooth muscle cells of arterial walls, caused by xenobiotics, as an alternative theory to the lipid theory in the development of atherosclerosis. However, the environmental contaminants may also influence serum lipoprotein levels.[56,57] Chlorinated hydrocarbons increase both total cholesterol and HDL cholesterol concentrations in experimental animals altering, at the same time, the apoprotein moieties.[35] These influences are probably mediated through the increased endogenous cholesterol synthesis possibly caused by the increased cytochrome P-450 content in the liver. However, phenobarbitone which is known as a potent cytochrome P-450 inducer, decreases serum cholesterol in experimental animals by lowering the cholesterol absorption.[58] Cigarette smoking is one of the most common environmental contaminants to which many persons are voluntarily exposed. Despite the fact that cigarette smoking may have numerous effects on man, its effects on serum lipoproteins may also be of importance. Cigarette smoking is regarded as one of the major risk factors for coronary heart disease. It may partly mediate its effects via changes in serum lipoproteins. A recent study showed that cigarette smoking decreased serum apoproteins A1 and A2, both present in HDL fraction of lipoproteins. This lipoprotein has a positive significance in the prevention of the development of coronary heart disease.[59]

REFERENCES

1. **Gotto, A. M., Jr., Shepherd, J., Scott, L. W., and Manis, E.,** Primary hyperlipoproteinemia and dietary management, in *Nutrition, Lipids, and Coronary Heart Disease,* Levy, R., Rifkind, B., Dennis, B., and Ernst, N., Eds., Raven Press, New York, 1979, 247.
2. **Moore, R. B., Long, J. M., Matts, J. P., Amplatz, K., Varco, R. L., Buchwald, H., and the Posch Group,** Plasma lipoproteins and coronary arteriography in subjects in the program on the surgical control of the hyperlipidemias, *Atherosclerosis,* 32, 101, 1979.
3. **Olsson, A. G.,** Signs of coronary atherosclerosis in apparently healthy men with different types of hyperlipoproteinemia, *Postgrad. Med. J.,* 51(Suppl. 8), 40, 1975.
4. **Carlson, L. A. and Bottiger, L. E.,** Ischaemia heart disease in relation to fasting values of plasma triglycerides and cholesterol, *Lancet,* 1, 865, 1972.
5. **Blackburn, H.,** Coronary disease prevention. Controversy and professional attitudes, *Adv. Cardiol.,* 20, 10, 1977.
6. **Wilhelmsen, L., Wedel, H., and Tibblin, G.,** Multivariate analysis of risk factors for coronary heart disease, *Circulation,* 48, 950, 1973.
7. **Fredrickson, D. S. and Lees, R. S.,** A system for phenotyping hyperlipoproteinemias, *Circulation,* 31, 321, 1965.
8. **Kuo, P. T.,** An assessment of hyperlipidemia (hyperlipoproteinemia) typing in the diagnosis and management of atherosclerosis, *Cardiovasc. Clin.,* 8, 33, 1977.
9. **Goldstein, J. L. and Brown, M. S.,** The low-density lipoprotein pathway and its relationship to atherosclerosis, *Ann. Rev. Biochem.,* 46, 897, 1977.
10. **Naito, C.,** Metabolism of lipoproteins, *Seikagaku,* 52, 102, 1980.
11. **Nikkilä, E. A., Huttunen, J. K., and Ehnholm, C.,** Low postheparin plasma hepatic lipase activity in familial type IIa hyperlipoproteinemia, *Ann. Clin. Res.,* 8, 63, 1976.
12. **Kritchevsky, D.,** Drug treatment of hyperlipideamias, in *Atherosclerosis,* Vol. 5, Gotto, A. M., Jr., Smith, L. C., and Allen, B., Eds., Springer-Verlag, New York, 1980, 71.
13. **Best, M. M. and Duncan, C. H.,** Hypoglyceridemic effect of ethyl-α-*p*-chlorphenoxyisobutyrate with and without androsterone, *Circulation,* 28, 690, 1963.
14. **Hellman, L., Zumoff, B., Kessler, G., Kara, E., Rubin, I. L., and Rosenfeld, R. S.,** Reduction of serum cholesterol and lipids by ethyl chlorophenoxyisobutyrate, *J. Atherosclerosis Res.,* 3, 454, 1963.
15. **Howad, R. P., Alaupouris, P., and Furman, R. H.,** Effects of chlorophenoxyisobutyrate and of chlorophenoxyisobutyrate plus androsterone on serum lipids and ultracentrifugally determined lipoproteins, *Circulation,* 28, 661, 1963.
16. **Macmillan, D. C., Oliver, M. F., Simpson, J. D., and Tothill, P.,** Effect of ethyl chlorophenoxyisobutyrate on weight, plasma volume, total body-water and free fatty acids, *Lancet,* 2, 924, 1965.
17. **Ryan, W. G. and Schwartz, T. B.,** The dynamics of triglyceride turnover: effects of Atromid-S, *J. Lab. Clin. Med.,* 64, 1001, 1964.
18. **Tolman, E. L., Tepperman, H. M., and Tepperman, J.,** Effect of ethyl p-chlorophenoxyisobutyrate on rat adipose lipoprotein lipase activity, *Am. J. Physiol.,* 218, 1313, 1970.
19. **Walton, K. W., Scott, P. J., Jones, J. V. Fletcher, R. F., and Whitehead, T.,** Studies on low density lipoprotein turnover in relation to atromid therapy, *J. Atherosclerosis Res.,* 3, 396, 1963.

20. **Decoopman, E., Sezille, G., Dewailly, P., Fruchart, J.-C., and Jaillard, J.,** Influence du clofibrate sur l'activité lipasique tissulaire chez le Rat, *Pathol. Biol.,* 24, 691, 1976.
21. **Nikkilä, E. A., Huttunen, J. K., and Ehnholm, C.,** Effect of clofibrate on postheparin plasma triglyceride lipase activities in patients with hypertriglyceridemia, *Metabolism,* 26, 179, 1977.
22. **Taylor, K. G., Holdworth, G., and Galton, D. J.,** Clofibrate increases lipoprotein-lipase activity in adipose tissue of hypertriglyceridaemic patients, *Lancet,* 2, 1106, 1977.
23. **Greten, H., Laible, V., Zipperle, G., and Augustin, J.,** Comparison of assay methods for selective measurement of plasma lipase, *Atherosclerosis,* 26, 563, 1977.
24. **Beaumont, J.-L. and Dachet, C.,** Binding to plasma lipoproteins of chlorophenoxyisobutyric, tibric and nicotinic acids and their esters. Its significance for the mechanism of lipid lowering by clofibrate and related drugs, *Atherosclerosis,* 25, 255, 1976.
25. **Goldberg, A. P., Applebaum-Bowden, D. M., Bierman, E. L., Hazzard, W. R., Haas, L. B., Sherrard, D. J., Brunzell, J. D., Huttunen, J. K., Ehnholm, C., and Nikkilä, E. A.,** Increase in lipoprotein lipase during clofibrate treatment of hypertriglyceridemia in patients on hemodialysis, *N. Engl. J. Med.,* 301, 1073, 1979.
26. **Kinnunen, P. K. J.,** High-density lipoprotein may not be antiatherogenic after all, *Lancet,* 2, 34, 1979.
27. **Kinnunen, P. K. J. and Virtanen, J.,** Mode of action of the hepatic endothelial lipase: recycling endocytosis via coated pits, in *Atherosclerosis,* Vol. 5, Gotto, A. M., Jr., Smith, L. C., and Allen, B., Springer-Verlag, New York, 1980, 383.
28. **Kuusi, T., Saarinen, P., and Nikkilä, E. A.,** Evidence for the role of hepatic endothelial lipase in the metabolism of plasma high density lipoprotein in man, *Atherosclerosis,* 36, 589, 1980.
29. **Nikkilä, E. A., Kuusi, T., Harno, K., Tikkanen, M., and Taskinen, M-R.,** Lipoprotein lipase and hepatic endothelial lipase are key enzymes in the metabolism of plasma high density lipoproteins, particularly of HDL_2, in *Atherosclerosis,* Vol. 5, Gotto, A. M., Jr., Smith, L. C., and Allen, B., Springer-Verlag, New York, 1980, 387.
30. **Vessby, B., Boberg, J., Gustafsson, J.-B., Karlström, B., Lithell, H., and Östlund-Lindqvist, A. M.,** Reduction of high density lipoprotein cholesterol and apolipoprotein A-1 concentrations by a lipid-lowering diet, *Atherosclerosis,* 35, 21, 1980.
31. **Vessby, B., Lithell, H., and Boberg, J.,** Supplementation with vitamin E in hyperlipidemic patients treated with diet and clofibrate. Effects on serum lipoprotein concentrations, plasma fatty acid composition and adipose tissue lipiprotein lipase activity, *Am. J. Clin. Nutr.,* 30, 547, 1977.
32. **Olsson, A. G.,** New drugs treatments in hyperlipidaemia, in *Atherosclerosis,* Vol. 5, Gotto, A. M., Jr., Smith, L. C., and Allen, B., Springer-Verlag, New York, 1980, 86.
33. **Arnold, A., McAuliff, J. P., Powers, L. G., Phillips, D. K., and Beyler, A. L.,** Ciprofibrate, a new orally effective hypolipemic agent, *Fed. Proc. Fed. Am. Soc. Exp. Biol.,* 38, 1149, 1979.
34. **Arnold, A., McAuliff, J. P., Powers, L. G., Phillips, D. K., and Beyler, A. L.,** The results of animal studies with ciprofibrate, a new orally effective hypolipidemic drug, *Atherosclerosis,* 32, 155, 1979.
35. **Sirtori, C. R., Franceschini, G., Sirtori, M., Gianfranceschi, G., and Poli, A.,** Apoprotein changes following treatments with hypolipidemic drugs, in *Atherosclerosis,* Vol. 5, Gotto, A. M., Jr., Smith, L. C. and Allen, B., Springer-Verlag, New York, 1980, 190.
36. **Sirtori, C. R., Catapano, A., Chiselli, G. C., Innocenti, A. L., and Rodriquez, J.,** Metformin: an antiatherosclerotic agent modifying very low density lipoproteins in rabbits, *Atherosclerosis,* 26, 78, 1977.
37. **Sirtori, C. R., Catapano, A., Ghiselli, G. C., Shore, B., and Shore, V. G.,** Effects of metaformin on lipoprotein composition in rabbits and man, *Prot. Biol. Fluids,* 25, 379, 1978.
38. **Goldberg, A. P., Chen, M., Brunzell, J. D., Bierman, E. L., and Porte, D., Jr.,** Treatment of hypertriglyceridemia with para-aminosalicylic acid-C: a possible mechanism of action, *Metabolism,* 27, 1648, 1978.
39. **Vessby, B., Lithell, H., Boberg, J., and Hellsing, K.,** Para-aminosalicylic acid as a lipid-lowering agent, *Clin. Pharmacol. Ther.,* 23, 651, 1978.
40. **Hunninghake, D. B.,** Drug treatment of type II hyperlipoproteinemia. Effects on plasma lipid and lipoprotein levels, in *Atherosclerosis,* Vol. 5, Gotto, A. M., Jr., Smith, L. C., and Allen, B., Springer-Verlag, New York, 1980, 74.
41. **West, R. J., Lloyd, J. K., and Leonard, J. V.,** Long-term follow-up of children with familial hyperchloresterolaemia treated with cholestyramine, *Lancet,* 2, 873, 1980.
42. **Kane, J. P., Tun, P., Malloy, M. J., and Haved, R. J.,** Synergism in drug treatment of familial hypercholesterolemia, in *Atherosclerosis,* Vol. 5, Gotto, A. M., Jr., Smith, L. C., and Allen, B., Eds., Springer-Verlag, New York, 1980, 78.

43. **Magide, A. A., Myant, N. B., and Reichl, D.,** The effect of nicotinic acid on the metabolism of the plasma lipoproteins of rhesus monkeys, *Atherosclerosis,* 21, 205, 1975.
44. **Miettinen, T. A.,** Effect of nicotinic acid on catabolism and synthesis of cholesterol in man, *Clin. Chim. Acta,* 20, 43, 1968.
45. **Shepherd, J., Packard, C. J., Patsch, J. R., Gotto, A., Jr., and Taunton, O. D.,** Effects of nicotinic acid therapy on plasma high density lipoprotein subfraction distribution and composition on apolipoprotein A metabolism, *J. Clin. Invest.,* 63, 858, 1979.
46. **Frisk-Holmberg, M., Jorfeldt, L., Juhlin-Dannfelt, A., and Karlsson, J.,** Metabolic changes in muscle on long-term alprenolol therapy, *Clin. Pharmacol. Ther.,* 26, 566, 1979.
47. **Nehcini, P.,** The influence of calcium on the antilipolytic action of propranolol, timolol, labetalol, verapamil, procaine, and papaverine in rat adipocytes, *Pharmacol. Res. Commun.,* 11, 785, 1979.
48. **Nilsson, A., Hansson, P-G., and Hökfelt, B.,** Effect of metoprolol on blood glycerol, free fatty acids, triglycerides and glucose in relation to plasmacatecholamines in hypertensive patients at rest and following submaximal work, *Eur. J. Clin. Pharmacol.,* 13, 5, 1978.
49. **Stankiewicz-Choroszucha, B. and Górski, J.,** Effect of substrate supply and beta-adrenergic blockade on heart glycogen and triglyceride utilization during exercise in the rat, *Eur. J. Appl. Physiol.,* 43, 11, 1980.
50. **Lehtonen, A. L. and Viikari, J.,** Long-term effect of sotalol on plasma lipids, *Clin. Sci.,* 57, 405, 1979.
51. **Streja, D. and Mymin, D.,** Effect of propranolol on HDL cholesterol concentrations, *Br. Med. J.,*2, 495, 1978.
52. **Miettinen, T. A., Huttunen, J. K., Ehnholm, E., Kumlin, T., Matilla, S., and Naukkarinen, V.,** Effect of long-term antihypertensive and hypolipidemic treatment on high density lipoprotein cholesterol and apolipoproteins A-1 and A-II, *Atherosclerosis,* 36, 249, 1980.
53. **Pagnan, A., Pessina, A. C., Hlede, M., Zanetti, G., and Dal Palú, C.,** Effects of labetalol on lipid and carbohydrate metabolism, *Pharmacol. Res. Commun.,* 11, 227, 1979.
54. **Lehtonen, A., Hietanen, E., Marniemi, J., and Peltonen, P.,** Lipoprotein lipase enzyme mediates the decrease of HDL cholesterol caused by beta blockers, *Ric. Sci. Educ. Perm.,* Suppl. 19, 245, 1981.
55. **Juchau, M. R., Bond, J. A., and Benditt, E. A.,** Aryl 4-mono-oxygenase and cytochrome P-450 in the aorta: possible role in atherosclerosis, *Proc. Natl. Acad. Sci. U.S.A.,* 73, 3723, 1976.
56. **Ishikawa, T. T., McMeely, S. S., Steiner, P. M., Glueck, C. J., Mellies, M., Gartside, P. S., and McMillin, C.,** Effects of chlorinated hydrocarbons on plasma α-lipoprotein cholesterol in rats, *Metabolism,* 27, 89, 1978.
57. **Reggiani, G.,** Medical problems raised by TCDD contamination in Seveso, Italy, *Arch. Toxicol.,* 40, 161, 1978.
58. **Byers, S. O., and Friedman, M.,** Phenobarbital as a hypocholesterolemic agent in the rat and rabbit, *Metabolism,* 25, 727, 1975.
59. **Berg, K., Börresen, A.-L., and Dahlen, G.,** Effect of smoking on serum levels of HDL apoproteins, *Atherosclerosis,* 34, 339, 1979.

Exertion, Lipoproteins and their Metabolism

Chapter 7

DYNAMIC EXERCISE TESTS

Rainer Rauramaa, Katriina Kukkonen, and Esko Länsimies

TABLE OF CONTENTS

I. INTRODUCTION

Exercise stress tests have several indications of which the most common is the diagnosis of coronary artery disease. In clinical practice traditional loading tests, with clinical observation, auscultation, and electrocardiographic monitoring can nowadays be completed by such modern methods as echocardiography and isotope scintigraphies.

Since physical inactivity is related to increased risk of coronary heart disease,[1–4] physical exercise during leisure is suggested to promote health. On the other hand, because sedentary middle-aged persons may have asymptomatic heart disease, all efforts should be taken to reveal such pathological cardiovascular processes which might lead to acute heart attack during exercise.[5–8] Therefore, before engaging in physical activity it is recommended that a thorough medical examination and laboratory tests including exercise tests are performed.[9]

The purpose of this review is to focus on some cardiovascular and metabolic aspects relevant to exercise stress testing in healthy persons as well as in some common patient groups before initiating physical training. The use of exercise stress testing in the diagnosis of myocardial ischemia and prognosis of coronary heart disease will not be dealt with in-depth here. In this respect the reader is referred to recent reviews by Faris et al.[10] and McHenry and Morris.[11]

II. CHARACTERISTICS AND OPTIONS IN EXERCISE TESTS

The objective of an exercise test is to examine the capacity of the oxygen delivery system and its limitations. Obviously, in established coronary heart disease the defect is central and myocardial oxygenation is hampered while in most noncoronary persons the disorder is peripheral, which means disturbances in oxygen utilization in the skeletal muscle.

In connection with exercise stress test ECG, heart rate, and blood pressure are the main cardiovascular functions registered. Myocardial oxygen consumption can be estimated using the so-called double product which means systolic blood pressure multiplied by heart rate.[12,13] Measurement of oxygen consumption ($\dot{V}O_2$) gives an estimate of the rate of aerobic energy metabolism in skeletal muscle. $\dot{V}O_2$ can be measured either directly (using, e.g., paramagnetic oxygen analysis[14]) or can be estimated from heart rate responses during increased loading on a bicycle.[15] For estimation of maximal $\dot{V}O_2$ on treadmill there are various mathematical formulas,[16] but such an indirect approach is less precise. In connection with direct measurement of oxygen consumption the METs (metabolic multiples) principle can be applied. One MET equals to the subject's oxygen consumption at rest (3.5 $m\ell/kg \times min$). This is useful when training prescription is based on data from exercise test.[17]

If only heart rate is used as an indicator of the intensity of exercise, it must be taken into consideration that mean maximal heart rate decreases with age and that the interindividual variation is also considerable.[15] In addition, in coronary heart disease, the increase in heart rate during exercise may be deficient and expected maximal target heart rate (related to age) will not always be achieved. Therefore performing a so-called submaximal exercise test with a prefixed target heart rate (75 to 90% of maximal) may be fallacious if only heart rate is followed.[18]

The question of submaximal vs. maximal exercise test has been debated.[19] Submaximal test has been suggested to be safer but the criteria of submaximality on the basis of heart rate are not clear as discussed above. It is not known to what proportion symptoms indicative of myocardial ischemia occur after the heart rate level of 85% of maximal heart rate has been reached if none have appeared before that.[20] In addition, the practical significance of ischemic myocardial changes occurring only in maximal exertion has been questioned.[21]

In Scandinavian countries, bicycle ergometer is most often used whereas in North America treadmill ergometer is the most popular. Testing protocols for bicycle ergometer include either work conducted, graded (each step lasting for 1 to 6 min[15]), or pulse conducted[22] exercise tests. For the various testing protocols on treadmill the reader is referred to several reviews.[19,23–25]

The advantages of bicycle ergometer include determination of work intensity without direct measurement of oxygen consumption and recording of ECG and blood pressure without great technical difficulties. However, riding a bicycle is a less physiological and less familiar type of physical activity for sedentary persons than walking on a treadmill. $\dot{V}O_2$ is greater on treadmill because of recruitment of larger muscle mass and less local fatigue than when bicycling. Performance on a motordriven treadmill can also be better controlled externally while bicycling is more patient-controlled, especially when using a mechanically braked bicycle.[19] For a more detailed discussion on conducting exercise tests in practice see, e.g., Faris et al.,[10] Mead,[16] and Sheffield and Roitman.[19]

III. CARDIORESPIRATORY RESPONSES TO EXERCISE IN HEALTHY AND DISEASED SUBJECTS

In *healthy* persons cardiorespiratory responses to graded exercise can be evaluated on the basis of several physiological and metabolic parameters. Increased heart rate as such or in relation to progressively increasing work load has most widely been used when assessing physical performance capacity. There is a nearly linear correlation between heart rate and work load. Heart rate is the most important factor in determining increased cardiac output during exercise since stroke volume increases only up to the level corresponding to 50% of maximal work load.[26] The rationale for using only heart rate in the assessment of physical working capacity is acceptable especially when serial testing is used but more pertinent information can be obtained by the direct measurement of oxygen consumption and carbon dioxide production. At present there are automated systems easy to operate and reliable in results. Respiratory quotient (the ratio of carbon dioxide produced and oxygen used) may also give valuable information of performance capacity during submaximal exercise levels. Ventilation and respiration rate increase linearily with oxygen consumption up to the submaximal level.

In *obese* persons oxygen consumption during submaximal exercise levels tends to be higher both on treadmill and bicycle as compared to normal weight subjects,[27,28] although only near or at the maximal level is the difference most distinct.[27] The higher oxygen consumption is simply due to greater body mass. In addition, blood pressure during submaximal exercise has been suggested to be higher in the obese.[28] When the obese are engaged in physical training programs, treadmill testing seems to be more sensitive in detecting functional improvement,[29] especially when combined with direct measurement of oxygen consumption.[30] Pulmonary function is usually hampered only in massive obesity. Even if there are no differences in respiratory capacity between the lean and the obese after training, increased aerobic power in the obese may be partially explained by the increased maximal ventilation.[31]

In *coronary heart disease* the typical features during exercise testing are symptoms (angina pectoris) and ST-T changes in ECG. Sometimes heart rate and blood pressure do not increase proportionally with increased loading compared to healthy individuals. Such a nonresponsive state may suggest insufficient myocardial function.[18] Choosing the criteria for significant ST-T changes in ECG either during exercise or immediately after it influences the sensitivity and specificity of the exercise test. A careful analysis of 33 studies comparing exercise test and coronary angiography showed a wide variation in these two parameters.[32] The little variation in specificity of a 0.2 mV S-T depression was offset by very low sensitivity. Such discrepancies were suggested to result from different methodological standards. Generally a negative exercise test is most reliable in asymptomatic subjects and in subjects with atypical chest pain, while a positive test is most valuable in patients with typical chest pain or after myocardial infarction.[33]

The prognostic value of exercise test in asymptomatic men proved useful only when combined with other physical findings such as calcification of coronary arteries[34] or conventional risk factors.[35] If one or more risk factors were present and two or more exertional risk predictors (chest pain during exercise, short duration of exercise, poor heart rate increase and ischemic S-T-depression) were present, the 18 times greater risk ratio in about 1% of healthy men was not altered by age.[35]

The significance of exercise testing soon after myocardial infarction has been evaluated in large series of patients recently in several studies.[36–41] In all these studies the group with increased mortality during 1 to 2 years after infarction could be described using either ST-segment depression or arrhythmias as the criteria of increased risk. No complications during the tests were reported. The subjects with low risk can be recommended for unsupervised exercise training at home.[42]

There are few reports concerning cardiorespiratory responses of *diabetics* during exercise tests. According to Hagan et al.[43] there are no significant differences in submaximal and maximal responses in insulin-dependent diabetic patients with a short duration of the disease as compared with normal subjects. Poor physical performance capacity has been reported in insulin-dependent diabetes of long duration[44] and in non-insulin dependent diabetes.[45] Direct determination of oxygen consumption in addition to the measurement of heart rate during exercise test is accentuated by the fact that heart responses may be variable in the diabetics due to autonomic neuropathy.[44,46,47]

IV. COMMENTS

Exercise tests are performed either for healthy, asymptomatic, or diseased persons. Traditional indication has been the diagnosis of coronary heart disease. Later the test has been applied for screening of healthy persons with or without risk factors before physical training. According to Chung[48] the indications in this respect apply for all men above 40 years of age and for all women above 50. In addition, if risk factors for coronary heart disease or symptoms exist, testing should be undertaken even at a younger age, always under a medical doctor's supervision.

REFERENCES

1. **Morris, J. N., Chave, S. P. W., Adam, C., Sirey, C., Epstein, L., and Sheehan, D. J.,** Vigorous exercise in leisure time and the incidence of coronary heart disease, *Lancet,* 1, 333, 1973.
2. **Morris, J. N., Everitt, M. G., Pollard, R., Chave, S. P. W., and Semmence, A. M.,** Vigorous exercise in leisure time: protection against coronary heart disease, *Lancet,* 2, 1207, 1980.
3. **Paffenbarger, R. S., Jr. and Hale, W. E.,** Work activity and coronary heart mortality, *N. Engl. J. Med.,* 292, 545, 1975.
4. **Paffenbarger, R. S., Jr., Wing, A. L., and Hyde, R. T.,** Physical activity as an index of heart attack risk in college alumni, *Am. J. Epidemiol.,* 108, 161, 1978.
5. **Gibbons, L. W., Cooper, K. H., Meyer, B. M., and Ellison, C.,** The acute cardiac risk of strenuous exercise, *JAMA,* 244, 1799, 1980.
6. **Koplan, J. P.,** Cardiovascular deaths while running, *JAMA,* 242, 2578, 1979.

7. **Thompson, P. D., Stern, M. P., Williams, P., Duncan, K., Haskell, W. L., and Wood, P. D.,** Death during jogging or running, *JAMA*, 242, 1265, 1979.
8. **Vuori, I., Mäkäräinen, M., and Jääskeläinen, A.,** Sudden death and physical activity, *Cardiology*, 63, 287, 1978.
9. Committee on Exercise and Physical Fitness, Evaluation for exercise participation. The apparently healthy individual, *JAMA*, 219, 900, 1972.
10. **Faris, J. V., McHenry, P. L., and Morris, S. N.,** Concepts and applications of treadmill exercise testing and the exercise electrocardiogram, *Am. Heart J.*, 95, 102, 1978.
11. **McHenry, P. L. and Morris, S. N.,** Exercise electrocardiography—current state of art, in *Advances in Electrocardiography*, Schlant, R. and Hurst, J. W., Eds., Grune & Stratton, New York, 1976, chap. 14.
12. **Amsterdam, E. A. and Mason, D. T.,** Exercise testing and indirect assessment of myocardial oxygen consumption in evaluation of angina pectoris, *Cardiology*, 62, 174, 1977.
13. **Dorossiev, D. L.,** Methodology of physical training, principles of training and exercise prescription, *Adv. Cardiol.*, 24, 67, 1978.
14. **Wilmore, J. H., Davis, J. A., and Norton, A. C.,** An automated system for assessing metabolic and respiratory function during exercise, *J. Appl. Physiol.*, 40, 619, 1976.
15. **Lange Andersen, K., Shephard, R. J., Denolin, H., Varnauskas, E., and Masironi, R.,** Fundamentals of Exercise Testing, World Health Organization, Geneva, 1971.
16. **Mead, W. F.,** Maximal exercise testing—Bruce protocol, *J. Fam. Pract.* 9, 479, 1980.
17. **Balke B.,** Prescribing physical activity, in *Sports Medicine*, Ryan, A. J. and Allman, F. L., Eds., Academic Press, New York, 1974, 505.
18. **Powles, S. C. P., Sutton, J. R., Wocks, J. R., Oldridge, N. B., and Jones, N. L.,** Reduced heart rate response to exercise in ischemic heart disease: the fallacy of the target heart rate in exercise testing, *Med. Sci. Sports*, 11, 227, 1979.
19. **Sheffield, L. T. and Roitman, D.,** Stress testing methodology, *Prog. Cardiovasc. Dis.*, 19, 33, 1976.
20. **McHenry, P. L.,** Risks of graded exercise testing, *Am. J. Cardiol.*, 39, 935, 1977.
21. **Davidson, R. M.,** Controversies in the use of exercise stress testing in the diagnosis and management of ischemic heart disease, *Cardiovasc. Clin.*, 8, 159, 1977.
22. **Arstila, M.,** Pulse-conducted triangular exercise-ECG test, *Acta Med. Scand. Suppl.*, 529, 1, 1972.
23. **Froelicher, V. F., Jr., Brammell, H., Davis, G., Noguera, I., Stewart, A., and Lancaster, M. C.,** A comparison of three maximal treadmill exercise protocols, *J. Appl. Physiol.*, 36, 720, 1974.
24. **Froelicher, V. F., Jr., Thompson, A. J., Noguera, I., Davis, G., Stewart, A. J., and Triebwasser, H. J.,** Prediction of maximal oxygen consumption. Comparison of the Bruce and Balke treadmill protocols, *Chest*, 68, 331, 1975.
25. **Pollock, M. L., Bohannon, R. L., Cooper, K. H., Ayres, J. J., Ward, A., White, S. R., and Linnerud, A. C.,** A comparative analysis of four protocols for maximal exercise stress testing, *Am. Heart J.*, 92, 39, 1976.
26. **Åstrand, P.-O. and Rodahl, K.,** *Textbook of Work Physiology. Physiological Bases of Exercise*, 2nd ed., McGraw-Hill, New York, 1977, chap. 6.
27. **Davies, C. T. M., Godfrey, S., Light, M., Sargeant, A. J., and Zeidifard, E.,** Cardiorespiratory responses to exercise in obese girls and young women, *J. Appl. Physiol.*, 38, 373, 1975.
28. **Bray, G. A., Whipp, B. J., Koyal, S. N., and Wasserman, K.,** Some respiratory and metabolic effects of exercise in moderately obese men, *Metabolism*, 26, 403, 1977.
29. **Kollias, J., Skinner, J. S., Barlett, H. L., Bergsteinova, B. S., and Buskirk, E. R.,** Cardiorespiratory responses of young overweight women to ergometry following modest weight reduction, *Arch. Environ. Health*, 27, 61, 1973.
30. **Freyschuss, U., and Melcher, A.,** Exercise expenditure in extreme obesity: influence of ergometry type and weight loss, *Scand. J. Clin. Lab. Invest.*, 38, 753, 1978.
31. **Kollias, J., Boileau, R. A., Barlett, H. L., and Buskirk, E. R.,** Pulmonary function and physical conditioning in lean and obese subjects, *Arch. Environ. Health*, 25, 146, 1972.
32. **Philbrick, J. T., Horwitz, R. I., and Feinstein, A. R.,** Methodological problems of exercise testing for coronary artery disease. Groups, analysis and bias, *Am. J. Cardiol.*, 46, 807, 1980.
33. **Buckendorf, W., Warren, S. E., and Vieweg, W. V. R.,** Suspected coronary artery disease among military aviation personnel, *Aviat. Space Environ. Med.*, 51, 1153, 1980.
34. **Langou, R. A., Huang, E. K., Kelley, M. J., and Cohen, L. S.,** Predictive accuracy of coronary artery calcification and abnormal exercise test for coronary artery disease in asymptomatic men, *Circulation*, 62, 1196, 1980.
35. **Bruce, R. T., DeRouen, T. A., and Hossack, K. F.,** Value of maximal exercise tests in risk assessment of primary coronary heart disease events in healthy men, *Am. J. Cardiol.*, 46, 371, 1980.

36. **Davidson, D. M. and DeBusk, R. F.,** Prognostic value of a single exercise test 3 weeks after uncomplicated myocardial infarction, *Circulation,* 61, 236, 1980.
37. **DeBusk, R. F., Davidson, D. M., Houston, N., and Fitzgerald, J.,** Serial ambulatory electrocardiography and treadmill exercise testing after uncomplicated myocardial infarction, *Am. J. Cardiol.,* 45, 547, 1980.
38. **Sami, M., Kraemer, H., and DeBusk, R. F.,** The prognostic significance of serial exercise testing after myocardial infarction, *Circulation,* 60, 1238, 1979.
39. **Smith, J. W., Dennis, C. A., Gassman, A., Gaines, J. A., Staman, M., Phibbs, B., and Marcus, F. I.,** Exercise testing three weeks after myocardial infarction, *Chest,* 75, 12, 1979.
40. **Starling, M. R., Crawford, M. H., Kennedy, G. T., and O'Rourke, R. A.,** Exercise testing early after myocardial infarction: predictive value of subsequent unstable angina and death, *Am. J. Cardiol.,* 46, 909, 1980.
41. **Theroux, P., Waters, D. D., Halphen, C., Debaisieux, J.-C., and Mizgala, H. F.,** Prognostic value of exercise testing soon after myocardial infarction, *N. Engl. J. Med.,* 301, 341, 1979.
42. **DeBusk, R. F., Houston, N., Haskell, W., Fry, G., and Parker, M.,** Exercise training soon after myocardial infarction, *Am. J. Cardiol.,* 44, 1223, 1979.
43. **Hagan, R. D., Marks, J. F., and Warren, P. A.,** Physiologic responses of juvenile-onset diabetic boys to muscular work, *Diabetes,* 28, 1114, 1979.
44. **Rubler, S. R., and Arvan, S. B.,** Exercise testing in young asymptomatic diabetic patients, *Angiology,* 27, 539, 1976.
45. **Ruderman, N. B., Ganda, O. P., and Johansen, K.,** The effect of physical training on glucose tolerance and plasma lipids in maturity onset diabetes, *Diabetes,* 28 (Suppl. 1), 89, 1979.
46. **Hilsted, J., Galbo, H., and Christensen, N. J.,** Impaired cardiovascular responses to graded exercise in diabetic autonomic neuropathy, *Diabetes,* 28, 313, 1979.
47. **Storstein, L. and Jervell, J.,** Response to bicycle exercise testing in long-standing juvenile diabetes, *Acta Med. Scand.,* 205, 227, 1979.
48. **Chung, E. K.,** Exercise ECG testing. Is it indicated for asymptomatic individuals before engaging in any exercise program?, *Arch. Intern. Med.,* 140, 895, 1980.

Chapter 8

EXERCISE TRAINING AND SERUM LIPIDS

Aapo Lehtonen

TABLE OF CONTENTS

I. TISSUE METABOLISM

A. Adaptation of Skeletal Muscle to Training

Endurance exercise results in increases in the maximum capacity to utilize O_2 made possible by adaptations in the skeletal muscles, in the cardiovascular system, and in the autonomic nervous system. Endurance training is accompanied by increased oxidative capacity of trained muscle, and by enhanced capacity to synthesize two potential intracellular energy stores, glycogen and triglyceride.[1] It appears that when skeletal muscle adapts to endurance exercise, it becomes more like cardiac muscle, in that its content of mitochondria and its capacity to generate ATP (adenosine triphosphate) from oxidation of pyruvate and fatty acids increases.[2] Skeletal muscle mitochondria also become more like heart mitochondria in their enzyme pattern. Different training regimens appear to be selective in their effects on muscle-fiber size,[3] but the relative frequency of each fiber type is probably inherited[4] and not influenced by training.[5] The high proportion of type I fibers in the muscles of marathon runners[6] is perhaps the result of self-selection for that type of event, since type I fibers are particularly suited to aerobic metabolism and are believed to be resistant to fatigue. Endurance training increases the cross-sectional area of type I fibers. With sprint or strength-training, both types of fiber increases in size, but the change is more pronounced in the type II fibers.[3]

Endurance exercise training results in an increase in maximum O_2 uptake capacity. Generally, the increase in maximum O_2 uptake capacity results equally from an increase in cardiac output secondary to a higher stroke volume and an increase in arteriovenous O_2 difference.[7] Thus, increased extraction of O_2 by the working muscles appears to play as important a role as increased cardiac output in bringing about the increase in maximum O_2 uptake capacity seen with endurance-exercise training.

It has been shown with serial muscle biopsies that during submaximal exercise of the same intensity, individuals deplete their muscle glycogen stores more slowly when they are trained than when they are untrained.[8,9] Individuals who have adapted to endurance exercise derive a greater percentage of their energy from oxidation of fatty acids and less from carbohydrate than do untrained individuals.[8–10] This is reflected in increased rate of conversion of ^{14}C-labeled long-chain fatty acids to CO_2.[10] It is well established that, during submaximal exercise, physically trained individuals have lower blood and muscle lactate levels than untrained individuals.

As a summary, skeletal muscle adapts to endurance exercise which increases the capacity for aerobic metabolism. This is reflected in an increase in the capacities of muscle to oxidize pyruvate and long-chain fatty acids. Underlying this increase in the ability to obtain energy by respiration is an increase in the levels of a number of mitochondrial enzymes. These include the enzymes involved in fatty acid oxidation, the enzymes of the citrate cycle, and the components of the respiration chain. As a result of these endurance-exercise induced biochemical adaptations, fast-twitch red (type IIA) and slow-twitch red (type I) skeletal muscle fibers become more like heart muscle in their enzyme patterns.

B. Energy Homeostasis in Exercise

Skeletal muscle mass is about 40% of body weight in normal individuals and accounts for 35 to 40% of total oxygen consumption in the resting state. During exercise, consumption of oxygen and metabolic fuels increases markedly to provide the adenosine triphosphate (ATP) necessary for the contractile process.

Adipose tissue is man's predominant energy store, because of the high caloric content and anhydrous nature of triglyceride. Normal man carries 10 to 20 kg of triglyceride, adequate calories for survival for 2 to 3 months. In contrast, available carbohydrate accounts for less than 2000 kcal: 350 g of muscle glycogen, 80 to 90 g of liver glycogen, and 20 g of glucose in extracellular water. Muscle lacks the enzyme, glucose-6-phosphatase and therefore muscle glycogen cannot contribute directly to the addition of glucose to the bloodstream. Muscle glycogen is thus of use in meeting the energy requirements of muscle fibers in which it is contained. Liver glycogen is a source of blood glucose and can be rapidly mobilized in response to exercise or hypoglycemia. Body protein in normal man approximates 10 to 11 kg of which one half to two thirds is in muscle, but protein in both muscle and other tissues serves a primary role for other than caloric storage.

C. Carbohydrate and Fatty Acid Utilization

Muscle glycogen is the major fuel consumed during the earliest phase at the work level over 50% of the maximal oxygen uptake.[12] If exercise continues for 10 to 40 min, glucose uptake by muscle rises to 7 to 20 times the basal level, depending on the intensity of the work performed.[13] Between approximately 65 and 85% of maximal oxygen uptake, work time being 45 to 200 min, glycogen in working muscle seems to be the limiting factor as the exhaustion point coincides with extremely low muscle glycogen values.[9] By 40 min of exercise, blood-borne glucose is responsible for 75 to 90% of the total carbohydrate consumed, reflecting a progressive decline in the availability of muscle glycogen as exercise continues.[13] Beyond 40 min of exercise, the rate of glucose utilization increases progressively until it peaks at 90 to 180 min and then declines slightly.[14] Fatty acid utilization continues to increase during prolonged exercise. Thus after 3 to 4 hr of continuous mild exercise, the relative contribution of fatty acids to total oxygen use is about twice that of carbohydrate.[14] The increase in uptake of free fatty acids is in direct proportion to their inflow. These findings indicate that, under normal circumstances, uptake of free fatty acids by exercising muscle is not primarily regulated by the muscle itself, but the external factors such as the rate at which fatty acids are mobilized from adipose tissue. The fuel utilization during prolonged periods of mild to moderate exercise may thus be characterized as a triphasic sequence in which muscle glycogen, blood glucose, and free fatty acids predominate as the major energy-yielding substrate.[15]

D. Liver Glycogen

It is known that during heavy exercise the glucose production from the liver is increased.[12,16] In the resting state, the rate of hepatic glucose production is approximately 150 mg/min; 75% is produced by glycogenolysis and the remainder by gluconeogenesis from alanine, lactate, pyruvate, and glycerol.[17] The increased glucose production during exercise results almost entirely from augmental glycogenolysis.[13] The total amount of glucose released from the liver during 40 min of heavy work is estimated to be 18 g—no more than 20 to 25% of the total hepatic glycogen stores in the postabsorptive state.[13] At the end of prolonged muscular work, appreciable amounts of glucose are produced by glycogenolysis in the liver. Values up to 1100 mg/min have been observed,[18] but still the blood sugar content tends to fall. The overall importance of gluconeogenesis in prolonged exercise is underscored by the estimation that 50 to 60 g of liver glycogen is mobilized in 4 hr of exercise—a 75% depletion of total glycogen stores of liver.[14]

E. Local Lipid Stores and Exercise

The decrease of muscle triglyceride concentration during exhaustive exercise has been estimated.[19] The amount of fatty acids and glucose that were oxidized was estimated on the basis of the respiratory quotient and the oxygen consumption during the exercise.

It was estimated that three fourths and one fourth of the fatty acids oxidized were derived from muscle triglycerides and plasma fatty acids, respectively. The preexercise concentration of triglycerides in the muscle was negatively correlated to the amount of work performed as was also the decrease in the muscle triglyceride concentration during the exercise. Well-trained subjects have on an average larger lipid stores than untrained subjects in muscles.[20–22] The triglyceride stores are larger in slow twitch than in fast twitch fibers.[21] During exercise the muscle triglycerides decreased more in slow twitch than in fast twitch fibers of rat quadriceps muscles.[23] The best trained persons have the largest triglyceride stores before the race and their triglyceride stores also decrease most during the race.[21] Although endurance training markedly enhances the capacity of muscles to metabolize fats, the factors that regulate the usage of lipids during prolonged exercise do not appear to be limited by the capacity of the fibers to oxidize fatty acids.[24] A factor central to the control of fatty acid oxidation seems to be the rate of free fatty acids (FFA) mobilization. Marked mobilization of fatty acids from adipose tissue deposits has been demonstrated after exercise.[25,26]

F. Hormones During Exercise

Hormonal responses are possible mediators of many of the metabolic changes observed after exercise. Hormonal changes during exercise include elevations of growth hormone, epinephrine, norepinephrine, cortisol,[27] and glucagon.[28] Catecholamines are promptly secreted during stressful exercise and are potent stimulators of hepatic glucose release and fatty acid mobilization in adipose tissue. Thus, along with growth hormone and glucagon, catecholamines promote the availability of adequate sources of oxidizable substrate for generation of energy. The decrease in insulin is noteworthy in exercise and hypoinsulinemia occurs despite a modest rise in blood glucose.[13,29] This suggests an inhibition in insulin secretion is probably mediated by the adrenergic nervous system. The finding that stimulation of glucose uptake by muscle occurs together with hypoinsulinemia during exercise indicates that enhancement of glucose consumption is not modulated by insulin.[13] Because glucose tolerance remains unimpaired, an increase in total body sensitivity to insulin characterizes the well-trained athlete.[29] Physical training increases insulin binding to monocytes in the resting state but results in a fall in insulin binding during acute exercise.[30] Changes in insulin binding in athletes thus may account for augmented insulin sensitivity at rest as well as a greater shift from carbohydrate to fat usage during exercise than is observed in untrained controls.[30] Well-trained long-distance runners are characterized by lowered fasting insulin levels as well as by a decreased insulin response to glucose challenge as compared with sedentary controls and as a sign of sympathoadrenal adaptation to endurance exercise training; the lowering of catecholamine levels is shown during exercise.[29–31]

II. EFFECTS OF PHYSICAL ACTIVITY ON SERUM LIPIDS

A. Serum Total Cholesterol

The relationship between physical activity and serum lipid levels, especially serum cholesterol, has not been definitively established. In many earlier studies where exercise was reported to exert a lowering effect on cholesterol levels, little or no information was available on the subjects' dietary habits and on the losses in body weight. Mann[32] demonstrated that doubling the caloric intake had no effect on serum lipid levels as long as the excess calories were dissipated through exercise. Cessation of exercise led to fat deposition and doubled the concentration of serum cholesterol. Weight reduction, brought about by reduced caloric intake, returned serum lipids to initial levels prior to the start of exercise. The results of this study make it difficult to differentiate whether the lowering of the serum lipids was due to exercise or indirectly by utilizing the excess calories. The decrease of serum cholesterol level has been shown in some studies.[33–35] Campbell[36] trained obese and control subjects at the same absolute work level by walking them on the treadmill 3 hr/week for 10 weeks. Significant decreases in serum cholesterol were reported for the clinically obese, but not for the controls. The obese group was the only group to record a weight loss due to the training program.

Golding[37] demonstrated the 30% decrease in serum cholesterol level of 4 overweight persons, who trained for 25 weeks, 1 hr/day, 5 days/week. The decrease of serum cholesterol level paralleled losses in body weight. Dalderup et al.[38] maintained body weight in several groups of males exercising 80 min a day 5 days/week, and reported a slight decrease in serum cholesterol (231 mg % to 221 mg %) after several weeks of training. The authors also reported an increase in body density. In a study on adult males, Milesis[39] reported a decrease in skinfold measurements while body weight remained constant and serum lipids decreased. Several reports indicate that significantly lower total cholesterol levels (compared to sedentary controls) are associated with vigorous activity, for instance, in middle-aged male[40] and female[41] long-distance runners, and in young elite long-distance runners.[41] Thompson et al.[42] found a decrease 7.6 to 10% in total cholesterol 4 to 66 hr after a 42 km footrace.

Many studies have not demonstrated a cholesterol-lowering effect of exercise. Prolonged, heavy exercise did not lower the serum cholesterol level in a person participating in a yearly ski racing.[43] Holloszy et al.,[44] studying the effects of a 6-month physical conditions program on serum lipids of middle-aged men, reported a nonsignificant increase from 241 to 254 mg % for serum cholesterol. The increase in serum cholesterol parelleled body weight increases owing to positive caloric balance. Lewis et al.[45] could not demonstrate a cholesterol-lowering effect of exercise and they suggest that when plasma cholesterol is within normal range, it is unlikely to be lowered significantly by increased physical activity and/or weight reduction. Goode et al.[46] found no significant decrease of serum cholesterol after exercise. No decrease of serum total cholesterol concentrations of 6 sedentary obese men was found after 16 weeks of vigorous walking 90 min, 5 days/week on a treadmill expending about 1100 kcal per session.[47] As a total cholesterol measurement contains within it the elements of a low density lipoprotein (LDL) cholesterol and a high density lipoprotein (HDL) cholesterol, the predictive importance of total cholesterol is probably considerably diminished, particularly in physically active populations.

B. Serum Triglycerides

Exercise has been shown to decrease the turbidity of alimentary lipemia[48] as well as to reduce the increase in plasma triglyceride concentration normally seen after a fatty meal.[49] Carlson and Mossfeldt[43] found a pronounced decrease in the concentration of serum triglycerides in persons participating in a 90 km ski race. In a study of Holloszy et al.,[44] triglyceride levels decreased significantly during a 6-month physical conditioning program. A triglyceride-lowering effect of an exercise training program has been demonstrated in several studies and the decrease of triglycerides has been between 25 and 60%.[39,50–53] Measurements of very low density lipoprotein (VLDL) cholesterol in male and female long-distance runners have confirmed that the triglyceride-rich VLDL is indeed at a very low level in these very active individuals.[40]

Kirkeby et al.[54] studied serum lipids of 26 men participating in a 90-km cross-country ski race, before, immediately after, and on the following days. Serum triglyceride level was significantly decreased 1 day and 2 days after the ski race and returned to the initial level 4 days after the race. Thompson et al.[42] found that serum triglycerides of male runners participating in a 42-km footrace were unchanged up to 4 hr after the race, but at 18, 42, and 66 hr mean reductions of 65, 39, and 32% were observed. Thus, the decreasing effect of physical activity on serum triglycerides has been documented both after heavy acute exercise and particularly after longer periods of increased physical exercise. It is also reported that plasma triglyceride concentration and age are positively correlated in sedentary subjects but not in athletes.[51] There are some studies which are not consistent with these reports of triglyceride-lowering effect of physical activity.[55,56]

Calvy et al.[55] reported that a high caloric intake (4500 kcal/day) coupled with daily strenuous physical activity resulted in no change in cholesterol levels or body weight in young male subjects. However, triglyceride levels significantly increased from 42 to 91 mg %. Mann et al.[56] also reported an increase in plasma triglyceride concentrations after 6 months of moderate exercise and attributed the rise to greater food consumption. Sedgwick et al.[57] enrolled 370 sedentary men aged 20 to 65 years in a physical training program. Reexamination 5 years later showed that on average the subjects had not changed significantly in serum cholesterol and triglyceride concentrations or in other known risk factors of coronary heart disease. Many studies, however, are consistent in their findings of low mean triglyceride levels in very physically active groups.

C. Free Fatty Acids

Previous studies have shown that lipid oxidation during muscular activity is positively related to the concentration of circulating free fatty acids (FFA).[58] Even untrained muscle demonstrates a marked increase in lipid oxidation when plasma FFA concentrations are elevated.[59] The entry of FFA into the blood from adipose tissue increases rapidly with leg exercise.[60] Endurance training markedly enhances the capacity of muscles to metabolize fats,[24,61] thus muscles of trained individuals are able to oxidize more fat at a given concentration than those of untrained individuals. Bransford and Howley[62] found that the average increase in plasma FFA during the 60 min bicycle test of women was 0.22 μmol/ℓ prior to a training program. After 4 weeks training program the same absolute work load resulted in an increased plasma FFA of only 0.10 μmol/ℓ. These results agree with previous work on men which reported significantly lower plasma FFA following training[63] or lower levels in trained as compared to untrained men.[64] The increased fat combustion during prolonged, strenuous exercise is reflected by the observation that the fatty acid composition of serum FFA changed markedly towards that of adipose tissue during the exercise.[54]

It is suggested that glucose availability reduces exercise-induced glucagon secretion and, possibly as a consequence, FFA mobilization.[65] The plasma levels of FFA after training are lower than in untrained persons[66,67] but also unchanged FFA concentrations after training have been reported.[51,68] The lumberjacks whose occupational physical activity is most vigorous had significantly lower serum free fatty acid concentrations than electricians.[69] The serum concentrations of fatty acids were highest in control groups and lowest in ice-hockey and football teams during the active playing season.[70] The endurance runners, however, had significantly higher serum free fatty acid levels than inactive controls.[52]

D. Lipoproteins

Elevated concentrations of serum cholesterol have been indicated as a primary factor in atherosclerosis, but also the manner in which cholesterol is distributed among the plasma lipoproteins is important.[71,72] It has been proposed that the atherosclerotic process might be more successfully prevented by increasing plasma HDL cholesterol, than by attempts to reduce plasma cholesterol.[72] With respect to evidence suggestive of the fact that reduced HDL-cholesterol and elevated LDL-cholesterol may predispose one to atherosclerosis, it seems logical that alteration of these levels should be beneficial in reducing the incidence of coronary heart disease. Therefore, there has been much interest in determining lipoprotein changes produced by increased physical activity.

E. Cross-Sectional Studies

Carlson and Mossfeldt[43] found decreased levels of very low density lipoprotein and high levels of high density lipoprotein cholesterol in skiiers participating in a 90 km ski racing. Wood et al.[39] determined plasma lipoproteins in 41 very active men (running >25 km/week for the previous year, mean age 47 years, range 35 to 59 years) and in a comparison group of men. The runners had significantly decreased plasma low-density lipoprotein (LDL) cholesterol (125 vs. 139 mg/100 mℓ) concentrations and a higher mean level of high-density lipoprotein (HDL) cholesterol (64 vs. 43 mg/100 mℓ) than the comparison group. These very active men exhibited a plasma lipoprotein profile resembling that of younger women. Enger et al.[73] studied the lipoprotein of 220 well-trained men, usually performing heavy exercise twice or more weekly during the winter season. The blood samples were taken a day before participation in a cross-country ski race. Trained men had significantly higher HDL-cholesterol and HDL/total cholesterol ratio than untrained men, but did not differ significantly from untrained women. HDL-cholesterol was significantly higher in skiers above 60 years than in skiers of younger age. Lehtonen and Viikari[52] determined the serum lipids of 23 regularly training men, over 30 years old, who trained more than 25 km/week running or skiing. The exercise increased serum HDL cholesterol concentration. There was a positive correlation between the amount of weekly exercise in kilometers and plasma HDL cholesterol level. Exercising more than 70 km/week increased plasma HDL concentration clearly above the normal level. Martin et al.[74] found that elite distance runners had significantly higher HDL cholesterol and lower LDL cholesterol concentration than nonrunners. The ratio of HDL to LDL cholesterol was significantly higher in runners. Hartung et al.[75] investigated the effect of diet on HDL cholesterol in 59 marathon runners, 85 joggers (running at least 3.2 km three times per week), and 74 inactive men. Their results suggest that HDL differences (marathon runners, 65 mg %, joggers 58 mg %, and inactive men 43 mg %) among the three groups were primarily the result of distance run, not dietary factors. Distance run was also the best predictor of the HDL: total cholesterol ratio and of total cholesterol (a negative correlation). The results of Adner and Castelli[76] indicated that distance runners had elevated HDL cholesterol level, but there was no difference in low-density lipoprotein-cholesterol levels in runners and in controls.

F. Acute Effects of Exercise

Chronic heavy endurance exercise successfully alters serum lipoproteins. Enger et al.[77] investigated serum lipoproteins after a single exposure to a 70 km cross-country ski race. HDL cholesterol increased by 12% of the prerace level immediately after the race and was still elevated 4 days after the race. LDL + VLDL cholesterol showed a tendency to decrease immediately after the race and was reduced by 17 and 11% of the prerace level on the following 2 days. Thompson et al.[42] studied the concentrations of serum lipoproteins in 12 trained male runners participating in a 42-km footrace. Serum total cholesterol and triglyceride concentration decreased after the race, but only small and transient increases in HDL cholesterol levels were noted after exercise. The results suggest that prolonged exercise has little effect on HDL mass.

G. Longitudinal Studies

Lopez et al.[78] imposed four daily 30 min sessions per week on young men. After 7 weeks of exercise, significant decrease in serum triglycerides and VLDL cholesterol levels were found. The value of LDL cholesterol decreased and a concomitant increase in HDL values was observed. No changes in body weight were reported. Webster et al.[79] studied middle-aged men training 45 min a day, 5 days a week over a period of 12 weeks. No changes were observed in serum cholesterol, LDL-cholesterol, or HDL/LDL-cholesterol ratio. Significantly lower concentrations of triglycerides and VLDL-cholesterol after the training program were found. There was a trend for higher levels of HDL cholesterol after the first 3 weeks of training. A controlled trial is reported on the effects of mild-to-moderate physical activity on serum lipoprotein.[80] The exercise group participated in a 4-month exercise program that consisted of 3 to 4 weekly sessions. The control group was advised to maintain their previous exercise habits. Serum triglycerides decreased significantly and HDL cholesterol increased significantly from 1.27 ± 0.04 to 1.41 ± 0.04 mmol/ℓ in the exercise group during the trial. The level of LDL cholesterol decreased in both groups.

The results of Shephard et al.[81] are contradictory with the results of Huttunen et al.[80] Shephard and collaborators[81] studied the serum lipid profile of 98 men and 158 women before and after introducing an employee fitness program. The sample sorted itself into 4 subgroups (nonparticipants, drop-outs, high adherents and low adherents). The high adherents attended two to three 30 min physical activity classes per week over the 6-month period progressing to 15 to 17 min of aerobic activity per session, with significant gains of predicted maximum oxygen intake and reductions of body fat. Nevertheless, there were few favorable changes in lipids. A decrease of total cholesterol was unrelated to the change of maximum oxygen intake, and occurred equally in nonparticipants and high adherents. The main response of HDL cholesterol was a decrease in nonparticipants rather than an increase in high adherents. The investigators conclude, that if exercise is to modify serum lipids, it must be more strenuous than the usual industrial fitness program. However, in the study of Leon et al.[47] in 1979, there were observed significant increases in the HDL cholesterol levels of 6 sedentary obese men after 16 weeks vigorous walking 90 min, 5 day/week, program. The high/low density lipoprotein ratio increased 25.9%. Percentage body fat decreased from 23.3 to 17.4.

Streja and Mymin[82] studied the effect of a 13-week exercise program on lipids of sedentary middle-aged men with coronary artery disease. The preponderant component of the exercise program was walking or slow jogging adjusted to maintain the heart rate for a period of 20 to 30 min between 70 and 85% of the maximum obtained during previous treadmill testing. The HDL cholesterol levels increased, but there were no changes in serum triglycerides or LDL cholesterol levels. In the training program of Fanell and Barboriak,[83] the data suggest that after an initial fall in plasma HDL-cholesterol at 2 weeks, HDL-cholesterol rises linearly with a significant difference between 2-week and 8-week values with endurance training. Plasma triglycerides did not begin to decrease until 4 weeks of training.

In summary, a common pattern of change in plasma lipoprotein concentrations runs through these reports of training studies. The results are consistent with the findings of the cross-sectional studies and strongly suggest that apparently desirable changes can be induced in initially sedentary people by exercise-training programs.

H. Work Activity

Several studies show that the distribution of plasma lipoprotein cholesterol may be influenced by the level of habitual physical activity in leisure time, but any reports about the effect of vigorous physical activity at work have not been published earlier. Lehtonen and Viikari[69] have compared the serum lipid concentrations of full-time lumberjacks (13 men) whose occupational physical activity is most vigorous with those of a group of ordinary electricians. HDL-cholesterol levels of lumberjacks were significantly higher than those of electricians, but there were no differences in (VLDL + LDL)-cholesterol values. Three overweight lumberjacks had the lowest HDL-cholesterol concentrations, but HDL-cholesterol levels of other lumberjacks were clearly higher than normal. The work of lumberjacks is known to require more energy than any other occupation. Recorded energy expenditure figures range from 4500 to 8000 kcal/day.[84] According to this study the effect of vigorous work activity on serum HDL-cholesterol and also on triglycerides is similar to that of exercise in leisure time.

I. Apoproteins

A change in HDL cholesterol may reflect only a change in the proportion of cholesterol in relation to the other lipoprotein components, i.e., the apoproteins, A-I and A-II, and phospholipids. As physical activity seems to cause higher HDL cholesterol levels it is of interest to know whether this increase could also be demonstrated for the apoproteins A-I and A-II or whether it is a result of increased cholesterol:protein ratio. Lehtonen et al.[85] studied the amounts of apoproteins A-I and A-II in 23 athletes running at an average of at least 25 km/week and in 15 age matched controls. The athletes had significantly higher apolipoprotein A-I concentrations than the controls. No significant difference was noted in apolipoprotein A-II concentration between the groups. In about half of the athletes, apolipoprotein A-I concentrations were above the upper limit of the controls. The composition of HDL in the athletes did not differ significantly from that of the control HDL as judged from the ratio of HDL-cholesterol to A-I and A-II. Thompson et al.[42] noted small increases in apoprotein A-I levels 5 min after a 42-km footrace. Compared to the 1-hr prerace sample, apoprotein A-I levels were lower at 1, 4, 18, 42, and 66 hr after exercise. Huttunen et al.[80] noted that the concentrations of apolipoprotein A-I stayed constant after a 4-month exercise program. The level of apolipoprotein A-II decreased in both exercise and control groups. Miller et al.[86] found a strong positive correlation between aerobic capacity and HDL-cholesterol in 11 healthy males of varying degrees of habitual physical activity. The trend for aerobic capacity and apoprotein A-I increase was not significant. The molar ratio of HDL-cholesterol to apoprotein A-I, which reflects the serum HDL_2/HDL_3 ratio, was even more strongly correlated with aerobic capacity than was HDL-cholesterol concentration. These findings suggest that physical training itself raises the plasma HDL-cholesterol level through an effect on the synthesis and/or catabolism of the HDL_2 subclass. Krauss et al.[87] also found the significantly higher HDL_2 concentrations in serum of long-distance runners than in sedentary persons; HDL_3 was not different. The simultaneous increase of HDL-cholesterol and apo A-I caused by a training program did not correlate to each other in the study of Kiens et al.[88]

J. Amount of Exercise Training

The results of many studies strongly support the idea that the effect of increased physical activity on plasma lipoproteins is advantageous. However, these findings are prominent only after a considerable amount of regular, vigorous exercise. Lehtonen and Viikari[52] found a positive correlation between the number of kilometers run per week and the plasma HDL cholesterol concentration. Although there was a tendency to a higher than average HDL cholesterol concentration in the whole group of runners, the concentrations were clearly higher than normal only when the amount of running exceeded 70 km/week. Because of the amount of physical exercise needed to cause prominent changes in plasma lipoproteins seems to be relatively high, the contradictory results of training programs are understandable. In some studies[78,80,82] the training program which consists of three to four weekly over 30 min sessions has increased serum HDL cholesterol significantly compared to the control group. The intensity of exercise in these studies was adjusted to maintain the heart rate between about 70 and 85% of the maximum. Webster et al.[79] and Shephard et al.[81] could not show significant changes in serum lipoproteins of persons participating in the 3 or 6 months training program. The amount and intensity of exercise was approximately similar as in these studies, which showed significant changes in serum lipid concentrations. Shephard and collaborators conclude that if exercise is to modify serum lipids it must be more strenuous than the usual industrial fitness program. This high exercise requirement agrees with a study of Mjøs et al.[90] where no relation was found between HDL cholesterol and the amount of physical activity, which was obtained through a questionnaire. However, in the study of Leon et al.,[47] there were observed significant increases in the serum HDL cholesterol levels of men after 16 weeks vigorous walking 90 min 5 days/week program.

K. Quality and Quantity of Exercise—the Age and Sex of Subjects

Leisure time activities provide for many people the only opportunity for physical exertion. As seen previously, cross-sectional studies report that dedicated runners[39,52,75] have relatively high concentrations of HDL-cholesterol as compared to predominantly inactive controls matched for age. The quantity of exercise in the groups investigated in these studies is very high. Active long-distance runners, joggers, and skiers have usually trained regularly several years, several time per week. In all these studies the persons investigated are middle-age (about 40 years old) men. The serum level of HDL-cholesterol in women is higher than in men. The effect of exercise on serum lipoproteins of women has been investigated in some studies. Nikkilä et al.[89] compared the serum lipids of six physically active females with a control group. The six females were long-distance or track-finding runners and their weekly running distance was 40 to 90 km. The average age of the females was 25 years. The runners had higher serum HDL-cholesterol level compared to inactive controls. The concentration of LDL or VLDL lipoproteins were not different. A study by Lewis et al.[45] showed that 22 obese women who participated in a 17-week physical conditioning program exhibited a significant increase in plasma HDL/LDL-cholesterol ratio. However, these changes were accompanied by a significant reduction in body weight. Wood et al.[40] determined the serum lipids of 43 female runners, aged 30 to 59. All had averaged at least 15 miles per week running. Average miles run per week were 31 ± 15. HDL-cholesterol was higher in female runners than controls (75 ± 14 vs. 56 ± 14 mg/100 mℓ), while low-density lipoprotein cholesterol was lower (113 ± 33 vs. 124 ± 34). These differences were statistically significant and only partially attributable to known factors other than high physical activity level.

The metabolism during endurance exercise is mostly aerobic. There are many popular kinds of sports where the anaerobic part of metabolism is more prominent. When soccer players and ice-hockey players were compared in model matches, it was found that the energy expenditure in soccer players was 0.18 kcal/min/kg whereas the respective figure of ice-hockey players was 0.43 kcal/min/kg.[91] The anaerobic part of the metabolism of ice-hockey players is approximately two thirds of the total energy expenditure.[92] Lehtonen and Viikari[70] determined serum lipids of top-class football and ice-hockey players during different training programs. The serum levels of total cholesterol, HDL-cholesterol, and the ratio of HDL-cholesterol to total cholesterol were increased in those football players having more aerobic exercise in their training program. This aerobic part consisted of running at a speed of 4 to 5 min/km which induced a pulse level of about 140 beats/min. The concentrations of HDL-cholesterol and the HDL-cholesterol to total cholesterol ratio were lowest in the ice-hockey players whose training is most anaerobic. The training program of ice-hockey players had been planned to maintain the pulse rate mostly above the level of 180 beats/min. Each of the actions lasted 30 to 90 min. The ice-hockey team trained seven times per week (about 10 hr/week). According to the results of this investigation the differences of serum lipids seem to be in connection with the kinds of sport that differ in their energy metabolism requirements. However, the age of athletes may be of importance. The mean age of football players was about 23 years and of ice-hockey players about 25 years. When Rönnemaa et al.[93] investigated the serum lipids of young men (16 to 20 years old), there were not significant differences in HDL-cholesterol levels of top-class endurance runners, sprinters, and controls with ordinary physical activity. The serum LDL-cholesterol concentration was lowest in sprinters. The lipid profile of young men is relatively advantageous and although the high amount of exercise seems to cause a tendency to higher HDL-cholesterol and lower LDL-cholesterol levels, the changes are small. The importance of the quality of exercise has also been shown in the study of Nikkilä et al.[89] The sprinters (mean age 21 years) had a relatively light endurance training program with only 20 to 35 km running distance per week. Their chief training consisted of athletic exercises of short duration. In sprinters the serum lipid and lipoprotein concentrations did not differ significantly from those of controls. The long-distance runners (the average age 25 years) had higher mean levels of HDL-cholesterol than the respective controls.

Tennis is also a very popular kind of sport in many countries. Vodak et al.[94] have evaluated plasma lipid and lipoprotein concentrations in middle-aged male (mean age 42 years) and female (mean age 32 years) tennis players. The males reported regular tennis play for an average 9.9 years and the females for an average 6.9 years. During the time of investigation the men played an average 6.7 hr/week; the women played an average 5.4 hr/week. When compared to a sedentary group matched for age, sex, and education, the tennis players exhibited similar plasma total cholesterol and LDL-cholesterol concentrations and significantly lower triglyceride and VLDL-cholesterol concentrations. Plasma HDL-cholesterol was significantly higher in the tennis players. This increase of HDL-cholesterol in tennis players was independent of other factors known to alter plasma HDL-cholesterol concentration (the age, relative weight, cigarette smoking, alcohol intake, and oral contraceptive use).

The long-term results of physical fitness programs, where the amount of exercise is relatively small, do not seem satisfactory according to the results of Sedgwick et al.[95] The initial fitness program comprised of twice-weekly, 1 hr sessions of calisthenics, interval running, and volleyball. Reexamination 5 years later showed that men who had remained active and therefore had a higher degree of fitness did not differ in risk factors such as serum cholesterol and triglyceride levels from men who had returned to sedentary habits. After 5 years of the initial program one third of the men were continuing with regular exercise.

L. Effects of Physical Exercise on Hyperlipoproteinemia

As described earlier, serum triglycerides can be reduced as a result of heavy exercise. This evidence suggests that exercise can possibly be used as an effective means of normalizing fasting triglyceride levels in patients with hypertriglyceridemia.

Oscai et al.[53] found that four successive days of exercise effectively reduced fasting serum triglycerides. Exercise was also effective in correcting type IV and type V hyperlipoproteinemia. The investigators indicate that exercise exerts a transient effect on serum lipids since fasting triglyceride gradually returned to baseline levels within 3 to 7 days. Lampman et al.[96,97] studied the effectiveness of physical training, dietary management, and dietary management plus physical training in a group of 46 males with type IV hyperlipoproteinemia before and after 6 weeks of treatment. All treatments were effective in reducing serum triglyceride levels in type IV individuals. Physical training in combination with type IV hyperlipoproteinemia diets effectively normalized serum lipids. Gyntelberg et al.[98] investigated five middle-aged, male subjects with type IV hyperlipoproteinemia. Exercise 30 min/day for 4 days induced a decrease in serum triglyceride. The decrease of serum triglyceride was limited to the VLDL fraction. Since dietary calories were increased to balance the energy expenditure due to exercise the authors concluded that the triglyceride lowering effect of exercise was not mediated by a negative caloric balance.

The results of these four studies support the effectiveness of physical training either independently or in conjunction with dietary management and weight reduction in controlling type IV or type V hyperlipoproteinemia.

M. Lipogenesis and Lipolysis

Physically, very active persons have a relatively low percentage of total body weight in the form of fat. The size of fat cells is smaller than in sedentary individuals.[99] The low degree of adiposity in physically active subjects may be responsible for part of the decreased triglyceride concentration, since a positive correlation has been reported between relative weight and plasma triglycerides. Wood et al.[40] have estimated by matching runners to controls on the basis of relative weight that about 50% of the difference in triglyceride level between runners and sedentary controls could be accounted for by the greater relative weight of the controls. The mechanism by which exercise decreases serum triglyceride concentration remains to be established; it may be reduced hepatic production of triglycerides, augmented peripheral use, or both. Diversion of glucose into muscle and away from the liver may contribute to the lipid-lowering effect of exercise and may account for the depletion of hepatic glycogen stores, which lasts for several days.[53] If hepatic glycogen content was critical to lipogenesis, this series of events might account for the sustained 3 to 7 day hypolipodemic response to exercise. Similarly, augmentation of muscle use of free fatty acids for oxidative energy[15] may reduce the amount of such acids available for hepatic conversion to triglycerides.

Marked mobilization of fatty acids from adipose tissue deposits has been demonstrated after exercise.[26] Long-distance runners have been shown to have higher levels of muscle and adipose tissue lipoprotein lipase (LPL) activity than sedentary controls.[89] Calculated lipoprotein lipase activity present in whole body adipose tissue and skeletal muscle indicated that total LPL activity was 1.5 to 2.3 times higher in the runners. The adaptive increase of LPL activity enables the tissue to take up circulating triglycerides more avidly than in an untrained state. Training has been shown to enhance the glyceride synthesis in muscle and adipose tissue of rats.[100] Thus all the metabolic changes that develop during training aim to an accelerated turnover of fat in adipose tissue, blood, and muscle. The increased LPL activity could form an explanation for the lower serum triglyceride levels found in well-trained subjects. Physical training seems to increase the insulin sensitivity of tissues.[101] This change might account for the increase of LPL activity. Serum HDL cholesterol level and lipoprotein lipase activity of adipose tissue correlated highly.[89] HDL receives lipid and protein material in the circulation. Since the degradation of triglyceride-rich particles is largely dependent on the activity of LPL at peripheral tissues the amount of surface material transferred to HDL in plasma compartment is evidently increased by an increase of tissue LPL activity. Nikkilä et al.[89] concluded that this concept could account for the good correlation between HDL cholesterol and adipose tissue LPL activity.

The increase in the activity of the enzyme LCAT (lecithin/cholesterol acyltransferase), which may be involved in the transfer of unesterified cholesterol from cells to "nascent" HDL, has been reported.[78] Apoprotein A-I, which is the predominant apolipoprotein of HDL, is an activator of LCAT reaction. The concentration of apoprotein A-I is increased in physically active persons.

In summary, the exact mechanism by which exercise influences the serum lipoprotein levels has not been clarified. However, the exercise-induced adaptations in tissues help to explain the biochemical phenomena responsible for the changes of lipoprotein metabolism.

N. Effect of Sera of Men with High Physical Activity on the Synthesis of DNA and Glycosaminoglycans by Cultured Human Aortic Smooth Muscle Cells

The mechanisms by which lipoproteins exert their effects on the arteries are largely unknown. It has been proposed that stimulation of smooth muscle cells in arterial intima and media is of primary importance in atherogenesis.[102] The stimulation involves an increased proliferation rate of the cells and an enhanced production of collagen and glycosaminoglycans (GAG). Several studies with animals have shown that hyperlipidemic serum can cause increased proliferation rate of arterial smooth muscle cells.[103–105] The concentration[106] and synthesis[107] of arterial wall sulfated GAGS is increased in early atherosclerosis of human aorta, and the sulfated GAGs/hyaluronic acid ratio is also increased in the early atherosclerosis of human coronary arteries.[108] This ratio increases in experimental atherosclerosis even before macroscopic signs of arterial injury.[109]

The results of Tammi et al.[110] show that the synthesis of sulfated GAGs by human aortic smooth muscle cells is stimulated less by sera with high HDL-cholesterol level from active runners than by other sera and that the concentrations of HDL-cholesterol in the serum are negatively correlated with the synthesis of sulfated GAGs. In the presence of sera from runners the ratio of sulfated GAGs to hyaluronic acid was smaller than in the presence of other sera. The metabolism of HDL and aortic GAGs may thus be interrelated in a way that counteracts the initiation of atherosclerosis. Similar results were obtained in the study of Tammi et al.[111] where the effect of sera from high density lipoproteinemic lumberjacks engaging in vigorous physical activity on the incorporation of (^{3}H) thymidine and the synthesis of GAG by human aortic smooth muscle cells in culture was measured. High density lipoproteinemic serum inhibited significantly the incorporation of thymidine. The serum inhibited the synthesis of GAG and the inhibition was more pronounced for sulfated GAGs than for hyaluronic acid. The inhibition was of the same magnitude for the subclasses (chondroitin, dermatan, and heparan sulfates) of sulfated GAGs studied.

In conclusion these results show, using an in vitro model, that high density lipoproteinemia due to high physical activity inhibits those reactions of aortic smooth muscle cultures that are regarded as basic events in the pathogenesis of atherosclerosis. This effect seems to be dependent on the platelet factor because it was not present when platelet poor plasma was used.[112] It is possible that the effect of high density lipoproteinemic sera is due to HDL per se and do not reflect the activity of some unknown inhibitors whose occurrence coincides with changes in the concentration of HDL. However, the sera from runners with normal HDL do not cause changes in the synthesis of GAGs or in the incorporation of (^{3}H) thymidine as compared to normolipidemic physically inactive controls.

REFERENCES

1. **Morgan, T. E., Cobb, L. A., Short, F. A., Ross, R., and Gunn, D. R.,** Effect of long-term exercise on human muscle mitochondria, in *Muscle Metabolism During Exercise,* Pernow, B. and Saltin, B., Eds., Plenum Press, New York, 1971, 87.
2. **Holloszy, J. O.,** Adaptations of muscular tissue to training, *Prog. Cardiovasc. Dis.*, 18, 445, 1976.
3. **Saltin, B., Nazar, K., Costill, D. L., Stein, E., Lansson, E., Essén, B., and Gollnick, P. D.,** The nature of the training response, peripheral and central adaptations to one-legged exercise, *Acta Physiol. Scand.*, 96, 289, 1976.
4. **Komi, P. V., Viitasalo, J. H. T., Havu, M., Thorstensson, A., Sjödin, B., and Karlsson, J.,** Skeletal muscle fibers and muscle enzyme activities in monozygous and dizygous twins of both sexes, *Acta Physiol. Scand.*, 100, 385, 1977.
5. **Dons, B., Bollerup, K., Bond-Petersen, F., and Hancke, S.,** The effect of weightlifting exercise related to muscle fibre composition and muscle cross-sectional area in humans, *Eur. J. Appl. Physiol.*, 40, 95, 1979.
6. **Costill, D. L., Fink, W. J., and Pollock, M. J.,** Muscle fibre composition and enzyme activities of elite distance runners, *Med. Sci. Sports,* 8, 96, 1976.
7. **Ekblom, B.,** Effect of physical training on oxygen transport system in man, *Acta Physiol. Scand. Suppl.*, 328, 1, 1969.
8. **Hermansen, L., Hultman, E., and Saltin, B.,** Muscle glycogen during prolonged severe exercise, *Acta Physiol. Scand.*, 71, 129, 1967.
9. **Saltin, B. and Karlsson, J.,** Muscle glycogen utilization during work of different intensities, in *Muscle Metabolism During Exercise,* Pernov, B. and Saltin, B., Eds., Plenum Press, New York, 1971, 289.
10. **Issekutz, B., Miller, H. J., and Rodahl, K.,** Lipid and carbohydrate metabolism during exercise, *Fed. Proc. Fed. Am. Soc. Exp. Biol.*, 25, 1415, 1966.
11. **Saltin, B. and Karlsson, J.,** Muscle ATP, CP and lactate during exercise after physical conditioning, in *Muscle Metabolism During Exercise,* Pernow, B. and Saltin, B., Eds., Plenum Press, New York, 1971, 395.
12. **Hultman, E.,** Studies on muscle metabolism of glycogen and active phosphate in man with special reference to exercise and diet, *Scand. J. Clin. Lab. Invest.*, 19(Suppl. 94), 1, 1967.
13. **Wahren, J., Felig, P., Ahlborg, G., and Jorfeldt, L.,** Glucose metabolism during leg exercise in man, *J. Clin. Invest.*, 50, 2715, 1971.
14. **Ahlborg, G., Felig, P., Hagenfeldt, R., and Wahren, J.,** Substrate turnover during prolonged exercise in man, *J. Clin. Invest.*, 53, 1080, 1974.
15. **Felig, P. and Wahren, J.,** Fuel homeostasis in exercise, *N. Engl. J. Med.*, 293, 1078, 1975.
16. **Rowell, L. B., Masoro, E. J., and Spencer, M. J.,** Splanchnic metabolism in exercising man, *J. Appl. Physiol.*, 20, 1032, 1965.
17. **Felig, P.,** The glucose-alanine cycle, *Metabolism,* 22, 179, 1973.
18. **Hultman, E. and Nilsson, L. H.,** Liver glycogen: effect of diet and exercise, in *Muscle Metabolism During Exercise,* Pernow, B. and Saltin, B., Eds., Plenum Press, New York, 1971, 143.
19. **Fröberg, S. O., Carlson, L. A., and Ekelund, L.-G.,** Local lipid stores and exercise, in *Muscle Metabolism During Exercise,* Pernow, B. and Saltin, B., Eds., Plenum Press, New York, 1971, 143.
20. **Hoppeler, H., Lüthi, P., Claasen, H., Weibel, E. R., and Howald, H.,** The ultrastructure of the normal human skeletal muscle. A morphometric analysis of untrained men, women and well-trained orienteers, *Pflügers Arch.*, 344, 217, 1973.
21. **Lithell, H., Örlander, J., Schele, R., Sjödin, B., and Karlson, J.,** Changes in lipoprotein-lipase activity and lipid stores in human skeletal muscle with prolonged heavy exercise, *Acta Physiol. Scand.*, 107, 257, 1979.
22. **Morgan, T. E., Short, F. A., and Cobb, L. A.,** Alterations in human skeletal muscle lipid composition and metabolism induced by physical conditioning, *Biochem. Exercise Med. Sport,* 3, 116, 1969.
23. **Reitman, J., Baldwin, K. M., and Holloszy, J. O.,** Intramuscular triglyceride utilization by red, white and intermediate muscle and heart during exhausting exercise, *Proc. Soc. Exp. Biol.*, 142, 628, 1973.
24. **Costill, D. L., Fink, W. J., Getchell, L. H., Ivy, J. L., and Witzmann, F. A.,** Lipid metabolism in skeletal muscle of endurance-trained males and females, *J. Appl. Physiol.*, 47, 787, 1979.
25. **Carlson, L. A.,** Lipid metabolism and muscular work, *Fed. Proc. Fed. Am. Soc. Exp. Biol.*, 26, 1755, 1967.
26. **Keul, J., Haralambie, G., and Trittin, G.,** Intermittent exercise: arterial lipid substrates and arteriovenous differences, *J. Appl. Physiol.*, 36, 159, 1974.

27. **Hartley, J. H., Mason, J. W., Hogan, R. P., Jones, L. G., Kotchen, T., Mongey, E., Wherry, F., Pennington, L., and Ricketts, P.,** Multiple hormonal responses to graded exercise in relation to physical training, *J. Appl. Physiol.,* 33, 602, 1972.
28. **Felig, P., Wahren, J., Hendler, R., and Ahlborg, G.,** Plasma glucagon levels in exercising man, *N. Engl. J. Med.,* 287, 184, 1972.
29. **Lohman, O., Liebold, F., Heilmann, W., Singer, H., and Pohl, A.,** Diminished insulin response in highly trained athletes, *Metab. Clin. Exp.,* 27, 521, 1978.
30. **Koivisto, V., Soman, V., Conrad, P., Hendler, R., and Nadel, E.,** Insulin binding to monocytes in trained athletes. Changes in the resting state and after exercise, *J. Clin. Invest.,* 64, 1011, 1979.
31. **Winder, W. W., Hagberg, J. M., Hickson, R. C., Ehsanli, A. A., and McLane, J. A.,** Time course of sympathoadrenal adaptation to endurance exercise training in man, *J. Appl. Physiol.,* 45, 370, 1978.
32. **Mann, G. V.,** Importance of caloric disposition in cholesterol and lipoprotein metabolism of human subjects, *Fed. Proc. Fed. Am. Soc. Exp. Physiol.,* 14, 422, 1955.
33. **Campbell, D. E.,** Influence of diet and physical activity on blood serum cholesterol on young men, *Am. J. Clin. Nutr.,* 18, 79, 1966.
34. **Cooper, K. H., Pollock, M. L., Martin, R. P., White, S. R., Linnerud, A. C., and Jackson, A.,** Physical fitness levels versus selected coronary risk factors, A cross-sectional study, *JAMA,* 236, 166, 1976.
35. **Mann, G. V., Garrett, H. L., Farhi, A., Murray, H., and Billings, F. T.,** Exercise to prevent coronary heart disease, *Am. J. Med.,* 46, 12, 1969.
36. **Campbell, D. E.,** Effect of controlled running on serum cholesterol of young adult males of varying morphological constitutions, *Res. Quart.,* 39, 47, 1967.
37. **Golding, L. A.,** Effects of physical training upon total serum cholesterol levels, *Res. Quart.,* 32, 499, 1961.
38. **Dalderup, L. M., deVoogd, N., Meyknecht, E. A., and de Hartog, C.,** The effects of increasing the daily physical activity on the serum cholesterol levels, *Nutr. Diet.,* 9, 112, 1967.
39. **Milesis, C. A.,** Effects of metered physical training on serum lipids of adult man, *J. Sport. Med.,* 14, 8, 1974.
40. **Wood, P. O., Haskell, W., Klein, H., Lewis, I., Stern, M. P., and Farguhar, J. W.,** The distribution of plasma lipoproteins in middle-aged male runners, *Metabolism,* 25, 1249, 1976.
41. **Wood, B. D., Haskell, W. L., Stern, M. P., Lewis, S., and Perry, C.,** Plasma lipoprotein distributions in male and female runners. The marathon: physiological, medical, epidemiological and psychological studies, *N.Y. Acad. Sci.,* 301, 748, 1977.
42. **Thompson, P. D., Cullinane, E., Henderson, L. O., and Herbert, P. N.,** Acute effects of prolonged exercise on serum lipids, *Metabolism,* 29, 662, 1980.
43. **Carlson, L. A. and Mossfeldt, F.,** Acute effects of prolonged, heavy exercise on the concentration of plasma lipids and lipoproteins in man, *Acta Physiol. Scand.,* 62, 51, 1964.
44. **Holloszy, J. O., Skinner, J. S., Toro, G., and Cureton, T. K.,** Effects of a six-month program for endurance exercise on serum lipids of middle-aged men, *Am. J. Cardiol.,* 14, 753, 1964.
45. **Lewis, S., Haskell, W. L., Wood, P. W., Mannoogian, N., Bailey, J. E., and Pereira, M.,** Effects of physical activity on weight reduction in obese middle-aged women, *Am. J. Clin. Nutr.,* 29, 151, 1976.
46. **Goode, R. C., First Brook, J. B., and Shepard, R. J.,** Effects of exercise and a cholesterol-free diet on human serum lipids, *Can. J. Physiol. Pharmacol.,* 44, 575, 1966.
47. **Leon, A. S., Conrad, J., Hunnighake, D. B., and Serfass, R.,** Effects of a vigorous walking program on body composition, and carbohydrate and lipid metabolism of obese young men, *Am. J. Clin. Nutr.,* 32, 1776, 1979.
48. **Cohen, H. and Goldberg, C.,** Effect of physical exercise on alimentary lipaemia, *Br. Med. J.,* 2, 509, 1960.
49. **Nikkilä, E. A. and Konttinen, A.,** Effect of physical activity on postrandial levels of fats in serum, *Lancet,* 1, 1151, 1962.
50. **Björntorp, P., Berchfold, P., Grimby, G., Lindholm, B., Sanne, H., Tibblin, G., Wilhelmsen, L.,** Effects of physical training on glucose tolerance, plasma insulin and lipids and on body composition in men after myocardial infarction, *Acta Med. Scand.,* 192, 439, 1972.
51. **Hurter, R., Peyman, M. A., Schwale, J., and Barnett, C. W. H.,** Some immediate and long-term effects of exercise on the plasma lipids, *Lancet,* 2, 671, 1972.
52. **Lehtonen, A. and Viikari, J.,** Serum triglycerides and cholesterol and serum high-density lipoprotein cholesterol in highly physically active men, *Acta Med. Scand.,* 204, 111, 1978.
53. **Oscai, L. B., Patterson, J. A., Bogard, D. L., Beck, R. J., and Rothermel, B. L.,** Normalization of serum triglycerides and lipoprotein electrophoretic patterns by exercise, *Am. J. Cardiol.,* 30, 775, 1972.

54. **Kirkeby, K., Strömme, S., Bjerkedal, J., Hertzenberg, L., and Refsum, H. E.,** Effects of prolonged, strenuous exercise on lipids and thyroxine in serum, *Acta Med. Scand.*, 202, 463, 1977.
55. **Calvy, G. L, Cady, L. D., Mufson, A., Nierman, J., and Gertler, M. M.,** The effect of strenuous exercise on serum lipids and enzymes, *JAMA*, 183, 1, 1963.
56. **Mann, G. V., Garrett, H. L., Farki, A., Murray, H., and Billings, F. T.,** Exercise to prevent coronary heart disease, *Am. J. Med.*, 46, 12, 1969.
57. **Sedgwick, A. W., Brotherhood, J. R., Harris-Davidson, A., Taplin, R. E., and Thomas, D. W.,** Long-term effects of physical training programme on risk factors for coronary heart disease in otherwise sedentary men, *Br. Med. J.*, 281, 7, 1980.
58. **Armstrong, D. T., Steele, R., Altschuler, N., Dunn, A., Bishop, J. S., and Debodo, R. C.,** Regulation of plasma free fatty acid turnover, *Am. J. Physiol.*, 201, 9, 1961.
59. **Costill, D. L., Coyle, E., Dalsky, G., Evans, W., Fink, W., and Jorpes, D.,** Effects of elevated plasma FFA and insulin on muscle glycogen usage during exercise, *J. Appl. Physiol.*, 43, 695, 1977.
60. **Havel, R. J., Naimark, A., and Borchgrevink, C. F.,** Turnover rate and oxidation of free fatty acids of blood plasma in man during exercise: studies during continuous infusion of palmitate-1-C^{14}, *J. Clin. Invest.*, 42, 1054, 1963.
61. **Mole, P. A., Oscai, L. B., and Holloszy, J. O.,** Adaptation of muscle to exercise. Increase in levels of palmityl CoA synthetase, carnitine palmityl transferase and palmityl CoA dehydrogenase and in the capacity of oxidize fatty acids, *J. Clin. Invest.*, 50, 2323, 1971.
62. **Bransford, D. R. and Howley, E. T.,** Effects of training on plasma FFA during exercise in women, *Eur. J. Appl. Physiol.*, 41, 151, 1979.
63. **Rennie, M. J. and Johnson, R. H.,** Alteration of metabolic and hormonal responses to exercise by physical training, *Eur. J. Appl. Physiol.*, 33, 215, 1974.
64. **Johnson, R. H., Walton, J. L., Krebs, H. A., and Williamson, D. H.,** Metabolic fuels during and after severe exercise in athletes and non-athletes, *Lancet*, 2, 452, 1969.
65. **Luyckx, A. S., Pirnay, F., and Lefebvre, P. J.,** Effect of glucose on plasma glucagon and free fatty acids during prolonged exercise, *Eur. J. Appl. Physiol.*, 39, 53, 1978.
66. **Bjernulf, A., Boberg, J., and Fröberg, S.,** Physical training after myocardial infarction. Metabolic effects during short and prolonged exercise before and after physical training in male patients after myocardial infarction, *Scand. J. Clin. Lab. Invest.*, 33, 173, 1974.
67. **LeBlanc, J., Boylay, M., Dulac, S., Jobin, M., Labrie, A., and Rousseau-Migneron, S.,** Metabolic and cardiovascular responses to norepinephrine in trained and nontrained human subjects, *J. Appl. Physiol.*, 42, 166, 1977.
68. **Keul, J., Doll, E., and Haralambie, G.,** Freie Fettsäuren, Glycerin und Triglyceride im arteriellen und femoral-venosen Blut vor und nach einem vierwöchigen Training, *Pflügers Arch.*, 316, 194, 1970.
69. **Lehtonen, A. and Viikari, J.,** The effect of physical activity at work on serum lipids with a special reference to serum high-density lipoprotein cholesterol, *Acta Physiol. Scand.*, 104, 117, 1978.
70. **Lehtonen, A. and Viikari, J.,** Serum lipids in soccer and ice-hockey players, *Metabolism*, 29, 36, 1980.
71. **Castelli, W. P., Doyle, J. T., Gordon, T., Hames, C. G., Hjortland, M. C., Hulley, C. B., Kagan, S., and Zukel, W. J.,** HDL cholesterol and other lipids in coronary heart disease, *Circulation*, 55, 767, 1977.
72. **Miller, G. J. and Miller, N. E.,** Plasma-high density lipoprotein concentration and development of ischaemic heart disease, *Lancet*, 1, 16, 1975.
73. **Enger, S., Herbjörsen, K., Erikssen, J., and Fretland, A.,** High density lipoproteins (HDL) and physical activity: the influence of physical exercise, age and smoking on HDL-cholesterol and the HDL-total cholesterol ratio, *Scand. J. Clin. Lab. Invest.*, 37, 251, 1977.
74. **Martin, R. P., Haskell, W. L., and Wood, P. D.,** Blood chemistry and lipid profiles of elite distance runners. The marathon: physiological, medical, epidemiological, and psychological studies, *N.Y. Acad. Sci.*, 301, 346, 1977.
75. **Hartung, G. H., Foreyt, J. P., Mitcell, R. E. Vlasek, J., and Gotto, A. M.,** Relation of diet to high-density lipoprotein cholesterol in middle-aged marathon runners, joggers and inactive men, *N. Engl. J. Med.*, 302, 357, 1980.
76. **Adner, M. M. and Castelli, W. P.,** Elevated high-density lipoprotein levels in marathon runners, *JAMA*, 243, 534, 1980.
77. **Enger, S. C., Strömme, S. B., and Refsum, H. E.,** High density lipoprotein cholesterol, total cholesterol and triglycerides in serum after a single exposure to prolonged heavy exercise, *Scand. J. Clin. Lab. Invest.*, 40, 341, 1980.
78. **Lopez-S., A., Vial, R., Balart, L., and Arroyave, G.,** Effect of exercise and physical fitness on serum lipids and lipoproteins, *Atherosclerosis*, 20, 1, 1974.
79. **Webster, W. A., Smith, J. C., LaRosa, R., Muesing, R., and Wilson, P.,** Effect of twelve weeks of jogging on serum lipoprotein of middle-aged men, *Med. Sci. Sports*, 10, 55, 1978.

80. **Huttunen, J. K., Länsimies, E., Voutilainen, E., Ehnholm, C., Hietanen, E., Penttilä, J., Siitonen, O., and Rauramaa, R.**, Effect of moderate physical exercise on serum lipoproteins. A controlled clinical trial with special reference to serum high-density lipoproteins, *Circulation*, 60, 1220, 1979.
81. **Shephard, R. J., Youldon, P. E., Cox, M., and West, C.**, Effects of 6-month industrial fitness programme on serum lipid concentrations, *Atherosclerosis*, 35, 277, 1980.
82. **Streja, D. and Mymin, D.**, Moderate exercise and high-density lipoprotein-cholesterol. Observations during a cardiac rehabilitation program, *JAMA*, 242, 2190, 1979.
83. **Fanell, P. A. and Barboriak, J.**, The time course of alterations in plasma lipid and lipoprotein concentrations during eight weeks of endurance training, *Atherosclerosis*, 37, 231, 1980.
84. **Karvonen, M. J., Rautaharju, P. M., Orma, E., Punsar, S., and Takkunen, J.**, Heart disease and employment, *J. Occup. Med.*, 3, 49, 1961.
85. **Lehtonen, A., Viikari, J., and Ehnholm, C.**, The effect of exercise on high density (HDL) lipoprotein apoproteins, *Acta Physiol. Scand.*, 106, 487, 1979.
86. **Miller, N. E., Rao, S., Lewis, B., Bjørsvik, G., Myhre, K., and Mjøs, O. D.**, High-density lipoprotein and physical activity, *Lancet*, 1, 111, 1979.
87. **Krauss, R. M., Lindgren, F. T., Wood, P. O., Haskell, W. L., Albers, J. J., and Cheung, M. C.**, Differential increases in plasma high density lipoprotein subfractions and apolipoproteins (Apo-LP) in runners, *Circulation*, 56, 4, 1977.
88. **Kiens, B., Jörgensen, J., Lewis, S., Jensen, G., Lithell, H., Vessby, B., Hoe, S., and Schnohr, P.**, Increased plasma HDL-cholesterol and apo A-I in sedentary middle-aged men after physical conditioning, *Eur. J. Clin. Invest.*, 10, 203, 1980.
89. **Nikkilä, E. A., Taskinen, M.-R., Rehunen, S., and Härkönen, M.**, Lipoprotein lipase activity in adipose tissue and skeletal muscle of runners: relation to serum lipoproteins, *Metabolism*, 27, 1661, 1978.
90. **Mjøs, O. D., Thelle, D. S., Førde, O. H., and Vik-Mo, H.**, Family study of high density lipoprotein cholesterol and the relation to age and sex, *Acta Med. Scand.*, 201, 323, 1977.
91. **Seliger, V.**, Energy metabolism in selected physical exercises, *Int. Z. Angew. Physiol.*, 25, 104, 1968.
92. **Seliger, V., Kostka, V., Grusova, D., Kovac, J., Machochova, J., Payer, M., Pribylova, A., and Urbankova, R.**, Energy expenditure and physical fitness of ice-hockey players, *Int. Z. Ankew, Physiol.*, 30, 283, 1972.
93. **Rönnemaa, T., Lehtonen, A., Järveläinen, H., Vihersaari, T., and Viikari, J.**, Serum lipids of young male athletes and the effect of their sera on cultured human aortic smooth muscle cells, *Scand. J. Sport Sci.*, 2, 23, 1980.
94. **Vodak, P. A., Wood, P. D., Haskell, W. L., and Williams, P. T.**, HDL-cholesterol and other plasma lipid and lipoprotein concentrations in middle-aged male and female tennis players, *Metabolism*, 29, 745, 1980.
95. **Sedgwick, A. W., Brotherhood, J. R., Harris-Davidson, A., Taplin, R. E., and Thomas, D. W.**, Long-term effects of physical training programme on risk factors for coronary heart disease in otherwise sedentary men, *Br. Med. J.*, 281, 7, 1980.
96. **Lampman, R., Santinga, J., Bassett, D., Mercer, N., Hock, D., Foss, M., and Block, W. D.**, Normalizing serum lipids in men with Type IV hyperprebetalipoproteinemia, *Med. Sci. Sports*, 10, 55, 1978.
97. **Lampman, R., Santinga, J., Hodge, M., Block, W. D., Flora, J., and Bassett, D.**, Comparative effects of physical training and diet in normalizing serum lipids in men with Type IV hyperlipoproteinemia, *Circulation*, 44, 652, 1977.
98. **Gyntelberg, F., Brennan, R., Holloszy, J. O., Schonfeld, G., Rennie, M. J., and Weidman, S. W.**, Plasma triglyceride lowering by exercise despite increased food intake in patients with Type IV hyperlipoproteinemia, *Am. J. Clin. Nutr.*, 30, 716, 1977.
99. **Björntorp, P., Holm, G., Jacobsson, B., Schiller-deJounge, K., Lundberg, P. A., Sjöström, L., Smith, U., and Sullivan, L.**, Physical training in human hyperplastic obesity. IV. Effects on the hormonal status, *Metabolism*, 26, 319, 1977.
100. **Askew, E. W., Huston, R. L., and Dohm, G. L.**, Effect of physical training on esterification of glycerol-3-phosphate by homogenates of liver, skeletal muscle, heart and adipose tissue of rats, **Metabolism**, 22, 473, 1973.
101. **Björntorp, P., Fahlen, M., Grimby, G., Gustafson, A., Holm, J., Rehnström, P., and Scherstin, T.**, Carbohydrate and lipid metabolism in middle-aged, physically well-trained men, *Metabolism*, 21, 1037, 1972.
102. **Ross, R. and Glomset, J. A.**, Atherosclerosis and the arterial smooth muscle cell, *Science*, 180, 1332, 1973.

103. **Fischer-Dzoga, K., Chen, R., and Wissler, R.,** Effects of serum lipoprotein on the morphology, growth and metabolism of arterial smooth muscle cells, *Adv. Exp. Med. Biol.*, 43, 299, 1974.
104. **Ledet, T., Fischer-Dzoga, K., and Wissler, R. W.,** Growth of rabbit aortic smooth muscle cells cultured in media containing diabetic and hyperlipidemic serum, *Diabetes*, 25, 207, 1976.
105. **Rönnemaa, T. and Doherty, N. S.,** Effect of serum and liver extracts from hypercholesterolemic rats on the synthesis of collagen by isolated aortar and cultured aortic smooth muscle cells, *Atherosclerosis*, 26, 261, 1977.
106. **Dalferes, E. R., Ruiz, H., Kumer, V., Radnakrishnamurthy, B., and Berenson, G. S.,** Acid mucopolysaccharides of fatty streakes in young human male aortas, *Atherosclerosis*, 13, 121, 1971.
107. **Sanwald, R., Ritz, E., and Wiese, G.,** Acid mucopolysaccharide metabolism in early atherosclerotic lesions, *Atherosclerosis*, 13, 247, 1971.
108. **Tammi, M., Seppälä, P. O., Lehtonen, A., and Möttönen, M.,** Connective tissue components in normal and atherosclerotic human coronary arteries, *Atherosclerosis*, 29, 191, 1978.
109. **Tammi, M., Rönnemaa, T., and Viikari, J.,** Rapid increase of glycosaminoglycans in the aorta of hypercholesterolemic rats; a negative correlation with plasma HDL concentration, *Acta Physiol. Scand.*, 105, 188, 1979.
110. **Tammi, M., Rönnemaa, T., Vihersaari, T., Saarni, H., Lehtonen, A., and Viikari, J.,** Effect of sera from hyperlipidemic subjects and high-density-lipoproteinemic runners on the synthesis of DNA and glycosaminoglycans by cultured human aortic smooth muscle cells, *Med. Biol.*, 57, 118, 1979.
111. **Tammi, M., Rönnemaa, T., Vihersaari, T., Lehtonen, A., and Viikari, J.,** High density lipoproteinemic due to vigorous physical work inhibits the incorporation of (^{3}H) thymidine and the synthesis of glycosaminoglycans by human aortic smooth muscle cells in culture, *Atherosclerosis*, 32, 23, 1979.
112. **Rönnemaa, T.,** Serum lipoprotein composition, platelet factor and arterial smooth muscle cells, *Acta Med. Scand.*, 642, 55, 1980.

Chapter 9

THE EFFECT OF CONTINUOUS, LONG-TERM EXERCISE ON SERUM LIPIDS

Jukka Marniemi

TABLE OF CONTENTS

I. INTRODUCTION

Changes in blood lipids caused by physical activity are interesting in regard to aspects of sports physiology, public health, and also of merely clinical practice. In analyzing blood lipids, the recent physical activity of the subject at present also has to be controlled as one possible determinant of the results obtained. Also recently, review articles dealing with one[1,2] or more[3–5] aspects of adaptation of lipid metabolism to physical exercise have been compiled. However, in these reviews, except that of Paul,[3] the effects of acute prolonged exercise have gained less attention, in general.

The changes in serum lipid levels caused by continuous, long-term exertion are mainly due to increased energy needs in working muscles.[3,6] The energy sources being located just in the strained muscle are muscular glycogen and triglyceride stores, containing rather high amounts of energy, approximately 6,500 kJ and 10,000 kJ, respectively.[6] Energy substrates in blood circulation are serum glucose, triglycerides, and free fatty acids. The amounts of blood glucose and fatty acids are rather limited, but nevertheless, blood-borne substrates are important sources of energy, because they are continuously produced and mobilized from other stores, e.g., liver and adipose tissue during exercise.

It is generally agreed that the longer the duration and the lower the intensity of the exercise is, the higher is the relative share of lipids as energy sources.[6,7] If, for instance, in the bicycle ergometer test the strain corresponds to 50% of the maximal oxygen uptake of the subject, after 1 hr exercise the share of carbohydrates from the oxidative metabolism is about half, that of free fatty acids 15 to 25%, and that of muscle triglycerides about 25%.[8] When the duration of the exercise is increasing, the proportion of lipids and especially that of free fatty acids can rise up to 90% from the total energy consumed.[7]

II. TRIGLYCERIDES

Changes in blood triglyceride content during prolonged exercise reflect the enhanced utilization of fatty acids as energy substrates. In this connection, lipoprotein lipase (LPL) enzyme located in the capillary endothelium is in the key position. This enzyme hydrolyzes blood triglycerides to free fatty acids and glycerol, when the fatty acids released can be utilized as energy sources in working muscles.[9] It has been shown recently that during prolonged, strenuous exercise of several hours skeletal muscle LPL activity is increased severalfold.[10,11] When the duration of the exercise is shorter, about 1 hr, the increase is less and difficult to be found.[12,13] The best trained subjects showed the least increase in LPL activity, and they also had the highest muscle triglyceride contents.[10] So it is obvious that the trained subject does not need to utilize extramuscular energy stores as early and not in the same magnitude as the untrained one does.

When the exertion lasts for 1 hr or longer, the serum triglyceride concentration usually decreases either immediately in the exercise situation[14–19] or 1 to 3 days after it.[16,18–23] Most of these kinds of studies have been carried out in the connection of long running or skiing competitions or hikes, which brings the difficulty in stabilizing and controlling the diet and drinking in often inadequate conditions. At present it can be concluded, however, that the postprandial rise in serum triglyceride levels is less during long-term exercise. On the question, whether this also holds true for fasting serum triglyceride values, contradictory results have been published.[11,15,24–27] However, in many of the studies where increased or unaltered triglyceride levels were obtained, the hemoconcentration factors during exercise have not been taken into account, as emphasized by Taskinen et al.[11] So it seems obvious that at least during the exercise of sufficient duration such as several hours, the fasting serum triglyceride levels are also decreased.[15,28] In hypercholesterolemic subjects the decrease apparently is attained by smaller effort.[29,30]

The major proportion (70 to 80%) of the decrease in triglycerides during prolonged exertion takes place in the very-low-density (VLDL) fraction of the lipoproteins where their fasting concentration is also highest.[15,28,29] Decrease, although smaller, has also been reported in low-density (LDL) and high-density (HDL) fractions.[15] This decrease correlates positively both with the total starting level of serum triglycerides and with their starting level in different lipoprotein fractions.[15]

In spite of the decrease in serum triglycerides described above, the relative share of blood triglycerides as energy sources for working muscle is to be considered rather small compared to the almost unlimited capacity of adipose tissue triglycerides in this respect.[3,31]

In addition to the increased utilization of fatty acids there also exists a possibility of decreased hepatic triglyceride synthesis in explaining the decrease of plasma triglycerides during exercise. Indeed, it has been shown recently that, in rats, chronic exercise,[32] and, in man, some other acute stress situations,[33] are connected with decreased hepatic VLDL and triglyceride secretion.

III. FREE FATTY ACIDS

It is well-known at present that during long-term exercise serum free fatty acids increase markedly, the increment being higher in longer exertion.[11,14,16,17,19,20,25] These fatty acids originate from the adipose tissue after the hydrolysis of triglycerides by hormone-sensitive lipase which is activated by enhanced catecholamines during prolonged exercise.[3] This increase in serum free fatty acids results in augmentation of fatty acid uptake by working muscles,[34] the uptake being directly proportional to the amount of fatty acids released from the adipose tissue.[35] The rise in serum fatty acids is higher when the exercise is accomplished under severe, such as hypoxic, conditions.[36,37] In fasting subjects the rise is smaller,[38,39] indicating the effective utilization of fatty acids from adipose tissue triglycerides as energy sources in fasting. The increase of free fatty acids is smaller also in trained subjects compared to untrained ones suggesting that in trained individuals a greater proportion of their energy requirements during moderate, prolonged exercise is obtained from intramuscular lipid stores.[40]

Because the free fatty acids in serum during prolonged exercise originate from the adipose tissue, the composition of serum free fatty acids changes towards the average fatty acid composition of the adipose tissue.[20,25,41,42] The fatty acid composition is modified also by the preferential muscular uptake of certain unsaturated fatty acids during exercise.[25,41–43]

Serum free fatty acids have recently aroused interest because artificially, with heparin-raised free fatty acids, have been reported to have a sparing effect on the muscle glycogen depletion during prolonged exercise.[44] However, in intramuscular triglycerides this sparing effect is not seen,[45] and the inhibition of the rise in serum free fatty acids did not impair the subjects from running 10 mi.[26]

IV. CHOLESTEROL

The alterations in the serum total cholesterol concentrations during the exercise of even relatively long duration are usually small.[11,15,25,29,46] However, if the serum cholesterol is followed for some days after a strenuous, long-term exertion, a significant reduction in total cholesterol is obtained.[18,20,23,47]

Just recently Enger et al.[18] have reported an interesting finding that serum HDL cholesterol rose immediately after a 70 km cross-country ski-race significantly (12%) above the starting level. The increment reached its maximum, 17% above the starting level, 1 day after the race, and 4 days after the race HDL cholesterol was still elevated. A decrease of about corresponding magnitude (17%) was found in VLDL and LDL cholesterol.[18] Thompson et al.[23] also found increases in serum HDL cholesterol and apoprotein A-1 levels after a 42 km footrace, but these rises were only small and transient. Parallel, although not significant, changes in HDL cholesterol have also been reported earlier to occur after prolonged exercise.[15,28,29] After a shorter strain, like 20-km running, this increase in HDL cholesterol is no more demonstrable.[11]

These studies clearly indicate that physical activity as such causes parallel changes in HDL cholesterol that previously have been found repeatedly in physically active, fit people.[1] HDL cholesterol is generally regarded as a protecting risk factor against the development of coronary heart disease.[48]

In the evaluation of the mechanisms involved in exercise-induced changes of cholesterol metabolism, it has been shown that already a relatively moderate exertion of 1 hr increases the biliary excretion of cholesterol into the duodenum.[49,50] So at least cholesterol output from the circulation is increased during prolonged exercise.

Compared to triglycerides, cholesterol, and free fatty acids, serum phospholipids during long-term exercise have been studied considerably less. However, during severe prolonged exertion serum phospholipids have also been shown to decrease considerably from the fasting values.[15,28] This decrease is seen in all lipoprotein fractions, but it is especially seen remarked in HDL[15] and LDL[28] fractions.

REFERENCES

1. **Wood, P. D. and Haskell, W. L.,** The effect of exercise on plasma high density lipoproteins, *Lipids,* 14, 417, 1979.
2. **Castelli, W. P.,** Exercise and high-density lipoproteins, *JAMA,* 242, 2217, 1979.
3. **Paul, P.,** Effects of long lasting physical exercise and training on lipid metabolism, in *Metabolic Adaptation to Prolonged Physical Exercise,* Howald, H. and Poortmans, J. R., Eds., Birkhäuser, Basel, 1975, 156.
4. **Parizkova, J.,** *Body Fat and Physical Fitness,* Martinus Nijhoff B. V./Medical Division, The Hague, 1977, 1.
5. **Moffatt, R. J. and Gilliam, T. B.,** Serum lipids and lipoproteins as affected by exercise: a review, *Artery,* 6, 1, 1979.
6. **Essén, B.,** Intramuscular substrate utilization during prolonged exercise, in *The Marathon: Physiological, Medical, Epidemiological, and Psychological Studies,* Vol. 301, Milvy, P., Ed., Annals of the New York Academy of Sciences, New York, 1977, 30.
7. **Gollnik, P. D.,** Free fatty acid turn-over and the availability of substrates as a limiting factor in prolonged exercise, in *The Marathon: Physiological, Medical, Epidemiological, and Psychological Studies,* Vol. 301, Milvy, P., Ed., Annals of the New York Academy of Sciences, New York, 1977, 64.
8. **Essén, B., Hagenfeldt, L., and Kaijser, L.,** Utilization of bloodborne and intramuscular substrates during continuous and intermittent exercise in man, *J. Physiol. London,* 265, 489, 1977.
9. **Robinson, D. S.,** The function of the plasma triglycerides in fatty acid transport, in *Comprehensive Biochemistry,* Vol. 18, Florkin, M. and Stotz, E. H., Eds., Elsevier, Amsterdam, 1970, 51.
10. **Lithell, H., Örlander, J., Schéle, R., Sjödin, B., and Karlsson, J.,** Changes in lipoprotein-lipase activity and lipid stores in human skeletal muscle with prolonged heavy exercise, *Acta Physiol. Scand.,* 107, 257, 1979.
11. **Taskinen, M.-R., Nikkilä, E. A., Rehunen, S., and Gordin, A.,** Effect of acute vigorous exercise on lipoprotein lipase activity of adipose tissue and skeletal muscle in physically active men, *Artery,* 6, 471, 1980.

12. **Lithell, H., Hellsing, K., Lundqvist, G., and Malmberg, P.,** Lipoprotein-lipase activity of human skeletal-muscle and adipose tissue after intensive physical exercise, *Acta Physiol. Scand.,* 105, 312, 1979.
13. **Marniemi, J., Peltonen, P., Vuori, I., and Hietanen, E.,** Triglyceride lipase activities of human postheparin plasma in obese and non-obese subjects in physical exercise, *Acta Physiol. Scand.,* 102 (Abstr.), 38, 1978.
14. **Nikkilä, E. A. and Konttinen, A.,** Effect of physical activity on postprandial levels of fats in serum, *Lancet,* 2, 1151, 1962.
15. **Carlson, L. A. and Mossfeldt, F.,** Acute effects of prolonged heavy exercise on the concentration of plasma lipids and lipoproteins in man, *Acta Physiol. Scand.,* 62, 61, 1964.
16. **Maron, M. B., Horvath, S. M., and Wilkerson, J. E.,** Blood biochemical alterations during recovery from competitive marathon running, *Eur. J. Appl. Physiol.,* 36, 231, 1977.
17. **Scheele, K., Herzog, W., Ritthaler, G., Wirth, A., and Weicker, H.,** Metabolic adaptation to prolonged exercise, *Eur. J. Appl. Physiol.,* 41, 101, 1979.
18. **Enger, S. C., Strømme, S. B., and Refsum, H. E.,** High density lipoprotein cholesterol, total cholesterol and triglycerides in serum after a single exposure to prolonged heavy exercise, *Scand. J. Clin. Lab. Invest.,* 40, 341, 1980.
19. **Haralambie, G. and Senser, L.,** Metabolic changes in man during long-distance swimming, *Eur. J. Appl. Physiol.,* 43, 115, 1980.
20. **Kirkeby, K., Strömme, S. B., Bjerkedal, I., Hertzenberg, L., and Refsum, H. E.,** Effects of prolonged, strenuous exercise on lipids and thyroxine in serum, *Acta Med. Scand.,* 202, 463, 1977.
21. **Holm, G., Jacobsson, B., Holm, J., Björntorp, P., and Smith U.,** Effects of submaximal physical exercise on adipose tissue metabolism in man, *Int. J. Obesity,* 1, 249, 1977.
22. **Holm, G., Björntorp, P., and Jagenburg, R.,** Carbohydrate, lipid, and amino acid metabolism following physical exercise in man, *J. Appl. Physiol.,* 45, 128, 1978.
23. **Thompson, P. D., Cullinane, E., Henderson, L. O., and Herbert, P. N.,** Acute effects of prolonged exercise on serum lipids, *Metabolism,* 29, 662, 1980.
24. **Sannerstedt, R., Sanbar, S. S., and Conway, J.,** Metabolic effects of exercise in patients with type IV hyperlipoproteinemia, *Am. J. Cardiol.,* 25, 642, 1970.
25. **Hurter, R., Swale, J., Peyman, M. A., and Barnett, C. W. H.,** Some immediate and long-term effects of exercise on the plasma lipids, *Lancet,* 2, 671, 1972.
26. **Norris, B., Schade, D. S., and Eaton, R. P.,** Effects to altered free fatty acid mobilization on the metabolic response to exercise, *J. Clin. Endocrinol. Metab.,* 46, 254, 1978.
27. **Soimajärvi, J., Karvinen, E., Vihko, V., and Rahkila, P.,** Effects of moderate, prolonged muscle work on certain energy metabolites in two groups of men differing in physical fitness, in *Abstr. Symp. Physical Performance and Muscle Metabolism,* Publications of the University of Kuopio, 1, 5, 1976.
28. **Carlson, L. and Fröberg, S. O.,** Blood lipid and glucose levels during a ten-day period of low-caloric intake and exercise in man, *Metabolism,* 16, 624, 1967.
29. **Gyntelberg, F., Brennan, R., Holloszy, J. O., Schonfeld, G., Rennie, M. J., and Weidman, S. W.,** Plasma triglyceride lowering by exercise despite increased food intake in patients with type IV hyperlipoproteinemia, *Am. J. Clin. Nutr.,* 30, 716, 1977.
30. **Giese, N. D., Nagle, F. J., Corliss, R. J., and Westgard, J. O.,** Diet and exercise: effects on serum lipids, *Circulation,* 50 (Suppl. 3), 175, 1974.
31. **Kaijser, L.,** Substrate utilization by skeletal muscle during prolonged physical exercise, in *Proc. Symp. Physical Performance and Muscle Metabolism,* Hänninen, O. and Harri, M., Eds., The Finnish Society for Research in Sport, Helsinki, 57, 1978, 159.
32. **Simonelli, C. and Eaton, R. P.,** Reduced triglyceride secretion: a metabolic consequence of chronic exercise, *Am. J. Physiol.,* 234, E221, 1978.
33. **Chait, A., Brunzell, J. D., Johnson, D. G., Benson, J. W., Palmer, J. P., Albers, J. J., Ensinck, J. W., and Bierman, E. L.,** Reduction of plasma triglyceride concentration by acute stress in man, *Metabolism,* 28, 553, 1979.
34. **Wahren, J., Hagenfeldt, L., and Felig, P.,** Glucose and free fatty acid utilization in exercise. Studies in normal and diabetic man, *Isr. J. Med. Sci.,* 11, 551, 1975.
35. **Ahlborg, G., Felig, P., and Hagenfeldt, L.,** Substrate turnover during prolonged exercise in man: splanchnic and leg metabolism of glucose, free fatty acids, and amino acids, *J. Clin. Invest.,* 53, 1080, 1974.
36. **Sutton, J. R.,** Effect of acute hypoxia on the hormonal response to exercise, *J. Appl. Physiol.,* 42, 587, 1977.
37. **Stock, M. J., Chapman, C., Stirling, J. L., and Campbell, I. T.,** Effects of exercise, altitude, and food on blood hormone and metabolite levels, *J. Appl. Physiol.,* 45, 350, 1978.
38. **Drenick, E. J., Fisler, J. S., Johnson, D. G., and McGhee, G.,** The effect of exercise on substrates and hormones during prolonged fasting, *Int. J. Obesity,* 1, 49, 1977.

39. **Pequignot, J. M., Peyrin, L., and Pérés, G.,** Catecholamine-fuel interrelationships during exercise in fasting men, *J. Appl. Physiol.*, 48, 109, 1980.
40. **Winder, W. W., Hickson, R. C., Hagberg, J. M., Ehsani, A. A., and McLane, J. A.,** Training-induced changes in hormonal and metabolic responses to submaximal exercise, *J. Appl. Physiol.*, 46, 766, 1979.
41. **Wirth, A., Neerman, G., Eckert, W., Heuck, C. C., and Weicker, H.,** Metabolic response to heavy physical exercise before and after a 3-month training period, *Eur. J. Appl. Physiol.*, 41, 51, 1979.
42. **Wood, P., Schlierf. G., and Kinsell, L.,** Plasma free oleic and palmitic acid levels during vigorous exercise, *Metabolism*, 14, 1095, 1965.
43. **Ledoux Peronnet, M., Ferguson, R., Allard, C., and Choquette, C.,** Lipemia and adipose tissue fatty acid composition in normal adults: effect of exercise and training, in Abstr. 4th Int. Symp. Biochemistry of Exercise, Brussels, 1979, 15.
44. **Costill, D. L., Coyle, E., Dalsky, G., Evans, W., Fink, W., and Hoopes, D.,** Effects of elevated plasma FFA and insulin on muscle glycogen usage during exercise, *J. Appl. Physiol.*, 43, 695, 1977.
45. **Stankiewicz-Choroszuchz, B. and Górski, J.,** Effect of decreased availability of substrates on intramuscular triglyceride utilization during exercise, *Eur. J. Appl. Physiol.*, 40, 27, 1978.
46. **Lavine, R. L., Lowenthal, D. T., Gellman, M. D., Klein, S., Vloedman, D., and Rose, L. I.,** Glucose, insulin and lipid parameters in 10,000 m running, *Eur. J. Appl. Physiol.*, 38, 301, 1978.
47. **Vuori, I., Marniemi, J., Rahkila, P., and Vainikka, M.,** The effect of a six-day ski-hike on plasma catecholamine concentrations and on their response to submaximal exercise, *Med. Biol.*, 57, 362, 1979.
48. **Gordon, R., Castelli, W. P., Hjortland, M. C., Kannel, W. B., and Dawber, T. R.,** High density lipoprotein as a protective factor against coronary heart disease—The Framingham Study, *Am. J. Med.*, 62, 707, 1977.
49. **Simko, V., Kelley, R. E., and Connell, A. M.,** The effect of short-term physical exercise on bile composition in human volunteers, *Gastroenterology*, 70, 938, 1976.
50. **Simko, V. and Kelley, R. E.,** Effect of physical exercise on bile and red blood cell lipids in humans, *Atherosclerosis*, 32, 423, 1979.

Chapter 10

EFFECT OF EXERCISE ON HORMONES REGULATING LIPOPROTEIN METABOLISM

Jukka Marniemi and Eino Hietanen

TABLE OF CONTENTS

I. INTRODUCTION

Numerous hormones have profound effects on the activities of enzymes regulating lipoprotein metabolism making it reasonable to find out the effect of exercise on the overall hormonal state. The relation of hormonal balance to physical exercise has already been the object of wide interest for a long time, and since the development of sensitive radioimmunoassays in the sixties, the number of studies on this subject has increased rapidly. A large amount of detailed reviews on the effect of exercise on one hormone or more have also been compiled. This chapter is more a short survey than a detailed description dealing only with those hormones considered important in the regulation of lipoprotein metabolism.

II. INSULIN

Plasma insulin concentrations can either increase, decrease, or stay unaltered during physical exercise, depending on the duration and intensity of the exercise and on the nutritional state and physical fitness of the subject. After a short-term maximal anaerobic exercise of a few minutes duration plasma insulin levels have been reported to be clearly (two to three times) elevated.[1–3] When the intensity is decreased and the duration of the exercise is lengthened, insulin levels are in general decreasing. Both increased[4,5] and decreased[6] insulin levels have been obtained after 10,000 m running. During prolonged aerobic exercise the decrease of plasma insulin is well-known.[7–11] The longer the exercise, the lower are insulin concentrations after the performance. This decrease is due to catecholamine mediated inhibition of insulin secretion from pancreatic B cells.[12–15] Thus, according to the present view, the inhibiting activity of the sympathetic nervous system rather than the changes in plasma glucose concentrations is the basic determinant of plasma insulin concentrations during exercise. When the exercise is extremely severe, such as leading to exhaustion[16–18] or being accomplished under hypoxic conditions,[19] slightly raised plasma insulin levels have been reported during the recovery period.

A decrease in fasting plasma insulin levels after physical training has repeatedly been documented in nonobese[20–23] and obese[24–26] subjects. Also the response of insulin to acute short-term and prolonged exercise is markedly modified by training. In the trained state similar, or even greater exertion, results in a smaller decrease of plasma insulin concentration than in the untrained state.[10,11,27–29] Even the insulin increase during the recovery phase after severe exercise is emphasized by training.[18] These changes are due to increased tissue sensitivity to insulin caused by physical training.[30–33] The basic alteration is mainly the increase in the concentration of insulin receptors.[31,33]

III. GLUCAGON

Blood glucagon levels are known to rise during physical exercise,[5–8,16,34] the rise being higher during prolonged submaximal than during short-term maximal exercise.[7–8] In general, the longer the duration of the exercise is, the higher is the rise in glucagon levels as also shown by comparing the effect of running 10 km, 25 km, or marathon.[6]

At present it is obvious that long-term physical training does not cause any marked changes in the fasting plasma glucagon concentrations.[28,29,35,36] However, according to some recent data the response of plasma glucagon to short-term[29] or prolonged, one bout of, exercise[28,29,36,37] is decreased considerably by aerobic training programs, the major proportion of the decrease in the response being seen already after 3 weeks of training.[36] In man, the increase in plasma glucagon levels during prolonged exercise is most obviously determined by direct effects of plasma glucose on pancreatic A cells and consequently on glucagon secretion.[9,13,15,38,39] However, the decreased plasma glucagon response to exercise after training is suggested to be due to a smaller catecholamine stimulus in the trained state.[29,36]

IV. CATECHOLAMINES

Closely connected to the exercise induced hormonal changes mentioned above is the now well-established increase in plasma catecholamines, norepinephrine and epinephrine, during exercise.[7,40–42] Usually norepinephrine concentration is affected more first, rises in epinephrine being detectable only after a severe or prolonged exercise.[8,9,40] However, also in plasma norepinephrine levels, more marked increases are brought about by postural changes from supine to standing, than by light or moderate exercise.[40] During prolonged endurance type exercise the plasma levels of both of these catecholamines continue to increase.[41] Recently the plasma concentration of the third catecholamine, dopamine, has also been shown to respond to maneuvers such as standing and physical exercise, causing changes in sympathetic activity in a similar way like epinephrine and norepinephrine.[43] However, compared to epinephrine and norepinephrine, the physiological significance of dopamine is poorly characterized.

According to the present view, the plasma norepinephrine concentration during physical exercise is mainly correlated to hemodynamic parameters and oxygen saturation, and it is not affected by changes in blood glucose concentration.[41,44] On the other hand, epinephrine concentrations are determined both by sympathetic nervous activity and partly by plasma glucose concentration during prolonged exercise.[44,45]

After a severe prolonged exercise like marathon running plasma epinephrine levels have been reported to be elevated still 1 day after the race, whereas those of norepinephrine had returned to the prerace level.[46] A check 3 days after a 6-day cross-country ski-hike (260 km) showed plasma epinephrine levels were, however, at the prehike level.[47] So it is obvious that the recovery of the sympathetic response to physical exercise is dependent on the severity of the exertion. It is now well-known that the sympathoadrenal response to short-term physical exercise is diminished after endurance training yielding lower increases in plasma epinephrine and norepinephrine levels.[10,28,36,37,48,49]

The major share of the decrement is already reached after 3 weeks training, after which the heart rate during exercise still continues to reduce.[36,49] So other factors than reduced sympathoadrenal response also seem to be involved in the training induced decrement in the exercise heart rate, and also in the more rapid recovery of the heart rate after the exercise in trained individuals.[50]

V. GLUCOCORTICOIDS AND ACTH

Conflicting results have been reported on the effect of both long-term and especially of short-term exercise on plasma glucocorticoids, mostly cortisol, during the last decade.[1,7,51,52] Earlier, mostly decreased[53,54] or unaltered[55–57] plasma cortisol levels were obtained after moderate or severe short-term exercise. However, many of these older results have to be interpreted with caution, because plasma cortisol levels are known to be already elevated sometimes before exercise apparently due to psychological stress factors involved in the test situation, causing an apparent diminution during exercise.[58,59] Vigorous effort of short duration most probably causes increased adrenocortical secretion of cortisol,[27,60] although very small or negligible increases have also been reported recently.[1,61]

During prolonged strenuous exercise the secretion and plasma levels of the "stress-hormone" cortisol is progressively increased[6,10,46,51,60,62–68] with the general rule that the greater the intensity of the exercise the greater the increase.[6,60] Under more stressful, such as hypoxic, conditions the increased secretion by exercise is still strengthened.[19] During the recovery phase after severe anaerobic running exertion the plasma cortisol concentrations return to the control day level in 3 hr,[60] whereas after a marathon race decreased cortisol levels have still been obtained 3 days after the race.[46]

Contradictory results exist in the literature on the effect of endurance training on cortisol secretion during acute exercise; both increased,[65] as well as unaltered[10,27,51,69] and lowered,[52,70–72] responses have been reported. Because in most convincing longitudinal studies no change in the cortisol levels were obtained in spite of clear increases in aerobic capacity,[10,27,69] it seems obvious that without the involvement of psychological anxiety factors the straight effect of aerobic training on adrenocortical response to acute exercise is rather small and at least difficult to demonstrate.

Also corticotropin (ACTH) secretion from the anterior pituitary is accelerated during prolonged severe exercise[65,73] being thus involved in the enhanced adrenocortical function during exercise.

VI. GROWTH HORMONE

Growth hormone (STH, somatotropin), also secreted from the anterior pituitary into the circulation, is found in plasma in clearly augmented concentrations after short-term aerobic[18,19,51,65,74–77] and anaerobic[1,74] exercise. The response observed depends on the age, physical fitness, and sex of the subjects being more likely in the elderly, unfit subjects, and females.[51] The intensity of the exertion also influences the response; the peak plasma growth hormone levels are higher[77] and are reached sooner[74] with more intense exercise. Under hypoxic conditions the increase in plasma growth hormone is emphasized.[19] In some studies a latent period of 10 to 20 min before any increase of plasma growth hormone[74,78] has been reported. This delay in the secretion of growth hormone apparently causes the peak plasma levels of the hormone often to be found after the short-term exercise during the recovery phase.[19,77]

The marked increase of plasma growth hormone during long-term aerobic exercise is now well-established.[10,37,51,75,79–85] When the intensity of the exercise is light or moderate, the response is more dependent on the psychological factors such as familiarity with the test situation. During moderate aerobic exercise the initial increase of growth hormone plasma level obtained during the first hour is often followed by a decrease towards the resting concentration when the exercise is continued.[80–82] With more severe exertion the secondary decline of growth hormone is more unusual, and the peak levels are higher.[51] However, new emotional stress factors such as those connected with extreme exhaustion[58,66,85] or competitive sports activity[75,80] confuse the interpretation of the results under some occasions and are certainly involved in the response obtained. The recovery of plasma growth hormone is rather rapid even after severe exercise under strong psychic stress during several days.[85] One night's sleep is enough to attain the resting growth hormone levels. However, higher nighttime levels have been reported after submaximal exercise during the day.[86] The recovery is slower when the exercise is accomplished under hypoxic conditions.[19]

It is now generally agreed that the plasma growth hormone response to a fixed level of exercise is affected by the fitness of the individual and thus by the relative exercise tolerance.[51] However, when the exercise is individually standardized to a constant percentage of aerobic performance capacity or the exercise is accomplished to exhaustion, the picture is more confused. In longitudinal studies both unchanged,[10,18] decreased,[27] and augmented[51] responses due to aerobic training programs have been reported. Most obviously these apparent discrepancies can be at least partly explained with the uncontrolled variation of psychological anxiety factors during the tests.

The basic mechanism of growth hormone release during exercise is still under discussion, although it seems likely that the hormone results from the adrenergic stimulation in the hypothalamus[83,87] and that humoral factors like serotonergic stimulation are also involved in the secretion.[88] Although decline in plasma free fatty acids increases the response of growth hormone during prolonged exercise,[84] it seems unlikely that the changes in plasma amino acids and free fatty acids could be responsible for the changed growth hormone levels, at least in the short-term exercise.[65,89]

VII. THYROID HORMONES

Thyroid function and the turn-over of thyroid hormones is also accelerated by exercise.[90–92] Short exercise of up to 20 min duration and of light to moderate intensity seems to result in no definite changes of free or total thyroxine (T_4) or triiodothyronine (T_3) concentrations in plasma.[93,94] With more severe and more prolonged exertion increased plasma levels have also been obtained frequently,[95–99] although not always.[93] When the exercise is continued for several days and is carried out under strong psychic stress, the initial increase of plasma T_3 and T_4 during the first day is followed by a gradual decrease reaching a minimum on the sixth day of the exertion.[85] Exercise at high altitude potentiates the rise in plasma thyroid hormones,[94] probably due to a shift from extrathyroidal tissue compartments to the extracellular one.

After severe prolonged exercise the recovery of plasma T_3 and T_4 is rather slow. Both increased[97] and reduced[85] concentrations are restored to the preexercise level in about 4 days. Regular physical training is accompanied by increased turnover of thyroid hormones, both secretion and degradation rates are accelerated,[90–92,100] resulting mostly in unchanged or slightly fallen plasma T_4 and T_3 concentrations.[92]

Recently, contrary to previous studies,[95–101] clearly raised plasma concentrations of thyrotropin (TSH) originating from the anterior pituitary have been demonstrated after prolonged[93,97] and even graded[93] exercise. This rise is most obviously due to enhanced secretion of TSH.[93] Thus it is very obvious that increased TSH concentrations at least partly account for the stimulated functions of the thyroid gland during physical training and exercise.

VIII. SEX HORMONES

Recently attention has also been focused on sex hormones, androgens, and estrogens together with pituitary FSH (follicle-stimulating hormone, follitropin) and LH (luteinizing hormone), when exercise induced hormonal changes are dealt with, at least partly due to the widespread use of anabolic steroids in sports.

Anaerobic exercise of very short duration like running about 15 sec even with maximal velocity does not seem to measurably change plasma testosterone and LH concentrations in spite of the significant increase in plasma androstenedione at the same time.[60,102] However, about 2 min long (3×300 m strenuous running) anaerobic exertion apparently is enough to increase plasma testosterone, androstenedione, LH, and estradiol levels.[1,61,102] Moderate aerobic physical strain obviously increases less plasma testosterone[58,60,93,102] and androstenedione[60] than does the strenuous one.[58,60,93,102] On the other hand, negative results with no change in blood testosterone during aerobic exercise have also been reported,[103–105] and Wilkerson et al.[106] have even suggested that the changes observed can be explained merely by decreases in plasma volume during exercise. However, this most obviously is not the case, because in plasma sex-hormone-binding globulin (SHBG) capacity no changes were obtained at the same time when testosterone concentration increased.[102]

An interesting phenomenon is that during prolonged exercise [85,93,107] or during the recovery after severe exercise,[1,60,62,102,108] plasma testosterone concentration is dramatically decreased. Most obviously psychic stress factors are involved at least in the decrease caused by prolonged exertion during several days under strenuous conditions, because one night's sleep partially restores the diminished plasma testosterone.[85]

After 6-months military training increased plasma total and free testosterone, LH, and androstenedione were found,[109] whereas after 8[108] and 15[110] weeks training no significant changes in serum testosterone levels were found. Also, when regularly training athletes were compared to untrained subjects no differences were found in plasma testosterone and SHBG.[102] Thus it seems that acute changes in testosterone levels found shortly after the beginning of the training gradually are smoothed when the training continues. In women, endurance training has recently been reported to cause alterations in the daily hormonal patterns of LH, FSH, progesterone, and estradiol, thus possibly interfering with the normal menstrual cycle.[111–113]

REFERENCES

1. **Adlercreutz, H., Härkönen, M., Kuoppasalmi, K., Kosunen, K., Näveri, H., and Rehunen, S.,** Physical activity and hormones, in *Advances in Cardiology,* Vol. 18, Manninen, V. and Halonen, P., Eds., S. Karger, Basel, 1976, 739.
2. **Hermansen, L., Pruett, E. D. R., Osnes, J. B., and Giere, F. A.,** Blood glucose and plasma insulin in response to maximal exercise and glucose infusion, *J. Appl. Physiol.,* 29, 13, 1970.
3. **Hermansen, L. and Vaage, O.,** Lactate disappearance and glycogen synthesis in human muscle after maximal exercise, *Am. J. Physiol.,* 2, E422, 1977.
4. **Lavine, R. L., Lowenthal, D. T., Gellman, M. D., Klein, S., Vloedman, D., and Rose, L. I.,** Glucose, insulin and lipid parameters in 10,000 m running, *Eur. J. Appl. Physiol.,* 38, 301, 1978.
5. **Lavine, R. L., Lowenthal, D. T., Gellman, M. D., Kline, S., Recant, R. L., and Rose, L. I.,** The effect of long-distance running on plasma immunoreactive glucagon levels, *Eur. J. Appl. Physiol.,* 43, 41, 1980.
6. **Scheele, K., Herzog, W., Ritthaler, G., Wirth, A. and Weicker, H.,** Metabolic adaptation to prolonged exercise, *Eur. J. Appl. Physiol.,* 41, 101, 1979.
7. **Galbo, H., Richter, E. A., Hilsted, J., Holst, J. J., Christensen, N. J., and Henriksson, J.,** Hormonal regulation during prolonged exercise, in *The Marathon: Physiological, Medical, Epidemiological, and Psychological Studies,* Vol. 301, Milvy, P., Ed., Annals of the New York Academy of Sciences, New York, 1977, 72.
8. **Galbo, H., Holst, J. J., and Christensen, N. J.,** Glucagon and plasma catecholamine responses to graded and prolonged exercise in man, *J. Appl. Physiol.,* 38, 70, 1975.
9. **Galbo, H., Holst, J. J., Christensen, N. J., and Hilsted, J.,** Glucagon and plasma catecholamines during beta-receptor blockade in exercising man, *J. Appl. Physiol.,* 40, 855, 1976.
10. **Hartley, L. H., Mason, J. W., Hogan, R. P., Jones, L. G., Kotchen, T. A., Mougey, E. H., Wherry, F. E., Pennington, L. L., and Ricketts, P. T.,** Multiple hormonal responses to prolonged exercise in relation to physical training, *J. Appl. Physiol.,* 33, 607, 1972.
11. **Rennie, M. J. and Johnson, R. H.,** Alteration of metabolic and hormonal responses to exercise by physical training, *Eur. J. Appl. Physiol.,* 33, 215, 1974.
12. **Brisson, G. R., Malaisse-Lagae, F., and Malaisse, W. J.,** Effect of phentolamine upon insulin secretion during exercise, *Diabetologia,* 7, 223, 1971.
13. **Galbo, H., Christensen, N. J., and Holst, J. J.,** Catecholamines and pancreatic hormones during autonomic blockade in exercising man, *Acta Physiol. Scand.,* 101, 428, 1977.
14. **Galbo, H., Richter, E. A., Christensen, N. J., and Holst, J. J.,** Sympathetic control of metabolic and hormonal response to exercise in rats, *Acta Physiol. Scand.,* 102, 441, 1978.

15. **Galbo, H., Holst, J. J., and Christensen, N. J.,** The effect of different diets and of insulin on the hormonal response to prolonged exercise, *Acta Physiol., Scand.,* 107, 19, 1979.
16. **Böttger, I., Schlein, E. M., Faloona, G. R., Knochel, J. P., and Unger, J. P.,** The effect of exercise on glucagon secretion, *J. Clin. Endocrinol.,* 35, 117, 1972.
17. **Drenick, E. J., Fisler, J. S., Johnson, D. G., and McGhee, G.,** The effect of exercise on substrates and hormones during prolonged fasting, *Int. J. Obesity,* 1, 49, 1977.
18. **Wirth, A., Neermann, G., Eckert, W., Heuck, C. C., and Weicker, H.,** Metabolic response to heavy physical exercise before and after a 3-month training period, *Eur. J. Appl. Physiol.,* 41, 51, 1979.
19. **Sutton, J. R.,** Effect of acute hypoxia on the hormonal response to exercise, *J. Appl. Physiol.,* 42, 587, 1977.
20. **Davidson, P. C., Shane, S. R., and Albrink, M. J.,** Decreased glucose tolerance following a physical conditioning program, *Circulation,* 33, 7, 1966.
21. **Björntorp, P., Fahlén, M., Grimby, G., Gustafson, A., Holm, J., Renström, P., and Scherstén, T.,** Carbohydrate and lipid metabolism in middle-aged, physically well-trained men, *Metabolism,* 21, 1037, 1972.
22. **Björntorp, P., Berchtold, P., Grimby, G., Lindholm, B., Sanne, H., Tibblin, G., and Wilhelmsen, L.,** Effects of physical training on glucose tolerance, plasma insulin and lipids and on body composition in men after myocardial infarction, *Acta Med. Scand.,* 192, 439, 1972.
23. **Streja, D. and Mymin, D.,** Moderate exercise and high-density lipoprotein cholesterol. Observations during a cardiac rehabilitation program, *JAMA,* 242, 2190, 1979.
24. **Björntorp, P.,** Exercise in the treatment of obesity, *Clin. Endocrinol. Metab.,* 5, 431, 1976.
25. **Holm, G., Sullivan, L., Jagenburg, R., and Björntorp, P.,** Effects of physical training and lean body mass on plasma amino acids in man, *J. Appl. Physiol.,* 45, 177, 1978.
26. **Krotkiewski, M., Mandroukas, K., Sjöström, L., Sullivan, L., Wetterqvist, H., and Björntorp, P.,** Effects of long-term physical training on body fat, metabolism, and blood pressure in obesity, *Metabolism,* 28, 650, 1979.
27. **Hartley, H. L., Mason, J. W., Hogan, R. P., Jones, L. G., Kotchen, T. A., Mougey, E. H., Wherry, F. E., Pennington, L. E., and Ricketts, P. T.,** Multiple hormonal responses to graded exercise in relation to physical training, *J. Appl. Physiol.,* 33, 602, 1972.
28. **Galbo, H., Richter, E. A., Holst, J. J., and Christensen, N. J.,** Diminished hormonal responses to exercise in trained rats, *J. Appl. Physiol.,* 43, 953, 1977.
29. **Gyntelberg, F., Rennie, M. J., Hickson, R. C., and Holloszy, J. O.,** Effect of training on the responses of plasma glucagon to exercise, *J. Appl. Physiol.,* 43, 302, 1977.
30. **Lohmann, D., Liebold, F., Heilmann, W., Senger, H., and Pohl, A.,** Diminished insulin response in highly trained athletes, *Metabolism,* 27, 521, 1978.
31. **Koivisto, V., Soman, V., Conrad, P., Hendler, R., Nadel, E., and Felig, P.,** Insulin binding to monocytes in trained athletes: changes in the resting state and after exercise, *J. Clin. Invest.,* 64, 1011, 1979.
32. **LeBlanc, J., Nadeau, A., Boylay, M., and Rousseau-Migneron, S.,** Effects of physical training and adipocity on glucose metabolism and ^{125}J-insulin binding, *J. Appl. Physiol.,* 46, 235, 1979.
33. **Soman, V. R., Koivisto, V. A., Deibert, D., Felig, P., and De Fronzo, R. A.,** Increased insulin sensitivity and insulin binding to monocytes after physical training, *N. Engl. J. Med.,* 301, 1200, 1979.
34. **Felig, P., Wahren, J., Hendler, R., and Ahlborg, G.,** Plasma glucagon levels in exercising man, *N. Engl. J. Med.,* 287, 184, 1972.
35. **Lampman, R. M., Santinga, J. T., Basset, D. R., Mercer, N., Block, W. D., Flora, J. D., Foss, M. L., and Thorland, W. G.,** Effectiveness of unsupervised and supervised high intensity physical training in normalizing serum lipids in men with type IV hyperlipoproteinemia, *Circulation,* 57, 172, 1978.
36. **Winder, W. W., Hickson, R. C., Hagberg, J. M., Ehsani, A. A., and McLane, J. A.,** Training-induced changes in hormonal and metabolic responses to submaximal exercise, *J. Appl. Physiol.,* 46, 766, 1979.
37. **Bloom, S. R., Johnson, R. H., Park, D. M., Rennie, M. J., and Sulaiman, W. R.,** Differences in the metabolic and hormonal response to exercise between racing cyclists and untrained individuals, *J. Physiol. London,* 258, 1, 1976.
38. **Luyckx, A. S., Pirnay, F., and Lefevbre, P. J.,** Effect of glucose on plasma glucagon and free fatty acids during prolonged exercise, *Eur. J. Appl. Physiol.,* 39, 53, 1978.
39. **Ahlborg, G. and Felig, G.,** Influence of glucose ingestion on fuel-hormone response during exercise, *J. Appl. Physiol.,* 41, 683, 1976.

40. **Christensen, N. J. and Brandsborg, O.,** The relationship between plasma catecholamine concentration and pulse rate during exercise and standing, *Eur. J. Clin. Invest.*, 3, 299, 1973.
41. **Christensen, N. J., Galbo, H., Hansen, J. F., Hesse, B., Richter, E. A., and Trap-Jensen, J.,** Catecholamines and exercise, *Diabetes*, 28, 58, 1979.
42. **Callingham, B. A.,** Catecholamines in blood, in *Handbook of Physiology*, Sec. 7, Vol. 6, Blaschko, H., Sayers, G., and Smith, A. D., Eds., Williams & Wilkins, Baltimore, 1975, 427.
43. **Van Loon, G. R., Schwartz, L., and Sole, M. J.,** Plasma dopamine responses to standing and exercise in man, *Life Sci.*, 24, 2273, 1979.
44. **Galbo, H., Christensen, N. J., and Holst, J. J., Glucose-induced decrease in glucagon and epinephrine responses to exercise in man,** *J. Appl. Physiol.*, 42, 525, 1977.
45. **Christensen, N. J., Alberti, K. G. M. M., and Brandsborg, G.,** Plasma catecholamines and blood substrate concentrations: studies in insulin induced hypoglycaemia and after adrenaline infusions, *Eur. J. Clin. Invest.*, 5, 415, 1975.
46. **Maron, M. B., Horvath, S. M., and Wilkerson, J. E.,** Blood biochemical alterations during recovery from competitive marathon running, *Eur. J. Appl. Physiol.*, 36, 231, 1977.
47. **Vuori, I., Marniemi, J., Rahkila, P., and Vainikka, M.,** The effect of a six-day ski-hike on plasma catecholamine concentrations and on their response to submaximal exercise, *Med. Biol.*, 57, 362, 1979.
48. **Cousineau, D., Ferguson, R. J., de Champlain, J., Gauthier, P., Cote, P., and Bourassa, M.,** Catecholamines in coronary sinus during exercise in man before and after training, *J. Appl. Physiol.*, 43, 801, 1977.
49. **Winder, W. W., Hagberg, J. M., Hickson, R. C., Ehsani, A. A., and McLane, J. A.,** Time course of sympathoadrenal adaptation to endurance exercise training in man, *J. Appl. Physiol.*, 45, 370, 1978.
50. **Hagberg, J. M., Hickson, R. C., McLane, J. A., Ehsani, A. A., and Winder, W.,** Disappearance of norepinephrine from the circulation following strenuous exercise, *J. Appl. Physiol.*, 47, 1311, 1979.
51. **Shephard, R. J. and Sidney, K. H.,** Effects of physical exercise on plasma growth hormone and cortisol levels in human subjects, in *Exercise and Sport Sciences Reviews*, Vol. 3, Wilmore, J. H. and Keogh, J. F., Eds., Academic Press, New York, 1975, 1.
52. **Tharp, G. D.,** The role of glucocorticoids in exercise, *Med. Sci. Sports*, 7, 6, 1975.
53. **Cornil, A., de Coster, A., Copinschi, G., and Franckson, J. R. M.,** Effect of muscular exercise on the plasma level of cortisol in man, *Acta Endocrinol. Copenhagen*, 48, 163, 1965.
54. **Raymond, L., Sode, J., and Tucci, J.,** Adrenocortical response to exercise, *Clin. Res.*, 17, 523, 1969.
55. **Metivier, G., Poortmans, J., Vanroux, R., Le Clercq, P., and Copinschi, G.,** Arterial blood plasma cortisol and human growth hormone changes in male trained subjects submitted to various physical work intensity levels, *Med. Sci. Sports*, 3, 9, 1971.
56. **Nakagawa, K. and Mashimo, K.,** Suppression of exercise-induced growth hormone release with dexamethasone, *Horm. Metab. Res.*, 5, 225, 1973.
57. **Moncloa, F., Carceler, A., and Beteta, L.,** Physical exercise, acid-base balance and adrenal function in newcomers to high altitude, *J. Appl. Physiol.*, 28, 151, 1970.
58. **Sutton, J. R., Coleman, M. J., Casey, J., and Lazarus, L.,** Androgen responses during physical exercise, *Br. Med. J.*, 1, 520, 1973.
59. **Sutton, J. R. and Casey, J. H.,** The adrecortical response to competitive athletics in veteran athletes, *J. Clin. Endocrinol.*, 40, 135, 1975.
60. **Kuoppasalmi, K., Näveri, H., Härkönen, M., and Adlercreutz, H.,** Plasma cortisol, androstenedione, testosterone and luteinizing hormone in running exercise of different intensities, *Scand. J. Clin. Lab. Invest.*, 40, 403, 1980.
61. **Kuoppasalmi, K., Näveri, H., Rehunen, S., Härkönen, M., and Adlercreutz, H.,** Effect of strenuous anaerobic running exercise on plasma growth hormone, cortisol, luteinizing hormone, testosterone, androstenedione, estrone and estradiol, *J. Steroid Biochem.*, 7, 823, 1976.
62. **Dessypris, A., Kuoppasalmi, K., and Adlercreutz, H.,** Plasma cortisol, testosterone, androstenedione and luteinizing hormone (LH) in a non-competitive marathon run, *J. Steroid Biochem.*, 7, 33, 1976.
63. **Davies, C. T. and Few, J. D.,** Effects of exercise on adrenocortical function, *J. Appl. Physiol.*, 35, 887, 1973.
64. **Few, J. D.,** Effect of exercise on the secretion and metabolism of cortisol in man, *J. Endocrinol.*, 62, 641, 1974.
65. **Sutton, J. R., Young, J. D., Lazarus, L., Hickie, J. B., and Maksvytis, J.,** The hormonal response to physical exercise, *Aust. Ann. Med.*, 18, 84, 1969.

66. **Sutton, J. R., Coleman, M. J., Millar, A. P., Lazarus, L., and Russo, P.**, The medical problems of mass participation in athletic competition, *Med. J. Aust.*, 2, 127, 1972.
67. **Newmark, S. R., Himathongkam, T., Martin, R. P., Cooper, K. H., and Rose, L. I.**, Adrenocortical response in marathon running, *J. Clin. Endocrinol.*, 42, 393, 1976.
68. **Cashmore, G. C., Davies, C. T. M., and Few, J. D.**, Relationship between increases in plasma cortisol concentration and rate of cortisol secretion during exercise in man, *J. Endocrinol.*, 72, 109, 1977.
69. **Amundsen, L. R. and Balke, B.**, Gluco-corticoid responses to acute and chronic physical exercise, *Med. Sci. Sports*, 5, 59, 1973.
70. **Frenkl, R., Csalay, L., and Csakvary, G.**, A study of the stress reaction elicited by muscular exertion in trained and untrained man and rats, *Acta Physiol.*, 36, 365, 1969.
71. **Frenkl, R., Csalay, L., and Csakvary, G.**, Further experimental results concerning the relationship of muscular exercise and adrenal function, *Endokrinologie*, 66, 285, 1975.
72. **Mikulaj, L., Komadel, L., Vigas, M., Kvetnansky, R., Starka, L., and Vencel, P.**, Some hormonal changes after different kinds of motor stress in trained and untrained young men, in *Metabolic Adaptation to Prolonged Physical Exercise*, Vol. 7, Howald, H., and Poortmans, J. R. Eds., Birkhäuser Verlag, Basel, 1975, 333.
73. **Viru, A., Smirnova, T., Tomson, K., and Matsin, T.**, Dynamics of Blood Levels of Pituitary Trophic Hormones during Prolonged Exercise, in Abstr. 4th Int. Symp. Biochemistry of Exercise, Brussels, June 19 to 22, 1979, 25.
74. **Buckler, J. M.**, Exercise as a screening test for growth hormone release, *Acta Endocrinol. Copenhagen*, 69, 765, 1972.
75. **Schalch, D. S.**, The influence of physical stress and exercise on growth hormone and insulin secretion in man, *J. Lab. Clin. Med.*, 69, 256, 1967.
76. **Glick, S. M., Roth, J., Yalow, R. S., and Berson, S. A.**, The regulation of growth hormone secretion, *Recent Prog. Horm. Res.*, 21, 241, 1965.
77. **Sutton, J. and Lazarus, L.**, Growth hormone in exercise: comparison of physiological and pharmacological stimuli, *J. Appl. Physiol.*, 41, 523, 1976.
78. **Tzankoff, S. P. and Robinson, S.**, Blood glucose, human growth hormone, immuno reactive insulin and lactic acid concentration in men during treadmill work, *Fed. Proc. Fed. Am. Soc. Exp. Biol.*, 33, 349, 1974.
79. **Roth, J., Glick, S. M., Yalow, R. S., and Berson, S. A.**, The influence of blood glucose on the plasma concentration of growth hormone, *Diabetes*, 13, 355, 1964.
80. **Hunter, W. M. and Greenwood, F. C.**, Studies on the secretion of human pituitary growth hormone, *Br. Med. J.*, i, 804, 1964.
81. **Hunter, W. M., Fonseka, C. C., and Passmore, R.**, Growth hormone: Important role in muscular exercise in adults, *Science*, 150, 1051, 1965.
82. **Hartog, M., Havel, R. J., Copinschi, G., Earll, J. M., and Ritchie, B. C.**, The relationship between changes in serum levels of growth hormone and mobilization of fat during exercise in man, *Q. J. Exp. Physiol. Cog. Med. Sci.*, 52, 86, 1967.
83. **Hansen, A. P.**, The effect of adrenergic receptor blockade on the exercise-induced serum growth hormone rise in normals and juvenile diabetics, *J. Clin. Endocrinol. Metab.*, 33, 807, 1971.
84. **Norris, B., Schade, D. S., and Eaton, R. P.**, Effects of free fatty acid mobilization on the metabolic response to exercise, *J. Clin. Endocrinol. Metab.*, 46, 254, 1978.
85. **Aakvaag, A., Sand, T., Opstad, P. K., and Fonnum, F.**, Hormonal changes in serum in young men during prolonged physical strain, *Eur. J. Appl. Physiol.*, 39, 283, 1978.
86. **Adamson, L., Hunter, W. M., Ogunremi, O. O., Oswald, I., and Percy-Robb, I. W.**, Growth hormone increase during sleep after daytime exercise, *J. Endocrinol.*, 62, 473, 1974.
87. **Sutton, J. R. and Lazarus, L.**, Effects of adrenergic blocking agents on growth hormone responses to physical exercise, *Horm. Metab. Res.*, 6, 428, 1974.
88. **Smythe, G. A. and Lazarus, L.**, Suppression of human growth hormone secretion by melatonin and cyproheptadine, *J. Clin. Invest.*, 54, 116, 1974.
89. **Hansen, A. P.**, The effect of intravenous infusion of lipids on the exercise-induced serum growth hormone rise in normals and juvenile diabetics, *Scand. J. Clin. Lab. Invest.*, 28, 207, 1971.
90. **Irvine, C. H. G.**, Effect of exercise on thyroxine degradation in athletes and non-athletes, *J. Clin. Endocrinol.*, 28, 942, 1968.
91. **Terjung, R. L. and Winder, W. W.**, Exercise and thyroid function, *Med. Sci. Sports*, 7, 20, 1975.
92. **Winder, W. W.**, Thyroid Hormones and Muscular Exercise, in Abstr. 4th Int. Symp. Biochemistry of Exercise, Brussels, June 19 to 22, 1979, 27.
93. **Galbo, H., Hummer, L., Petersen, I. B., Christensen, N. J., and Bie, N.**, Thyroid and testicular hormone responses to graded and prolonged exercise in man, *Eur. J. Appl. Physiol.*, 36, 101, 1977.

94. **Stock, M. J., Chapman, C., Stirling, J. L., and Campbell, J. T.,** Effects of exercise, altitude, and food on blood hormone and metabolite level, *J. Appl. Physiol.*, 45, 345, 1978.
95. **Terjung, R. L. and Tipton, C. M.,** Plasma thyroxine and thyroid-stimulating hormone levels during submaximal exercise in humans, *Am. J. Physiol.*, 220, 1840, 1971.
96. **Kirkeby, K., Strömme, S. B., Bjerkedal, I., Hertzenberg, L., and Refsum, H. E.,** Effects of prolonged, strenuous exercise on lipids and thyroxine in serum, *Acta Med. Scand.*, 202, 463, 1977.
97. **Refsum, H. E. and Strömme, S. B.,** Serum thyroxine, triiodothyronine and thyroid stimulating hormone after prolonged heavy exercise, *Scand. J. Clin. Lab. Invest.*, 39, 455, 1979.
98. **Caralis, E. L. and Davis, P. J.,** Serum total and free thyroxine and triiodothyronine during dynamic muscular exercise in man, *Am. J. Physiol.*, 233, E 115, 1977.
99. **Berchtold, P., Berger, M., Herrman, J., Rudorff, K., Zimmermann, H., and Krüskemper, H. L.,** Thyroid hormones and TSH during physical exercise in healthy and diabetic subjects, *Eur. J. Clin. Invest.*, 7, 222, 1977.
100. **Balsam, A. and Leppo, L. E.,** Effect of physical training on the metabolism of thyroid hormones in man, *J. Appl. Physiol.*, 38, 212, 1975.
101. **Federspil, G., Franchimont, P., and Hazee-Hagelstein, M. T.,** Serum TSH and prolactin levels during prolonged muscular exercise, *Horm. Metab. Res.*, 8, 323, 1976.
102. **Kuoppasalmi, K.,** Plasma testosterone and sex-hormone-binding globulin capacity in physical exercise, *Scand. J. Clin. Lab. Invest.*, 40, 411, 1980.
103. **Carstensen, H., Amér, B., Amér, I., and Wide, L.,** The postoperative decrease of plasma testosterone in man, after major surgery, in relation to plasma FSH and LH, *J. Steroid Biochem.*, 4, 45, 1973.
104. **Lamb, D. R.,** Androgens and exercise, *Med. Sci. Sports*, 7, 1, 1975.
105. **Brisson, G. R., Volle, M. A., Hélie, R., Lefrancois, C., de Carufel, D., Carpentier, J. P., Brault, J., Audet, A., and Desharnais, M.,** Blood Testosterone (bound & free), Δ^4-Androstenedione, FSH, LH and PRL in Trained Male Athletes Submitted to Light (55%), Medium (70%) and Heavy (85% of VO_2 max) Bicycle Work Loads, in Abstr. 4th Int. Symp. Biochemistry of Exercise, Brussels, June 19 to 22, 1979, 5.
106. **Wilkerson, J. E., Horvath, S. M., and Gutin, B.,** Plasma testosterone during treadmill exercise, *J. Appl. Physiol.*, 49, 249, 1980.
107. **Aakvaag, A., Bentdal, Ø., Quigstad, K., Walstad, P., Rønningen, H., and Fonnum, F.,** Testosterone and testosterone binding globulin (TeBG) in young men during prolonged stress, *Int. J. Androl.*, 1, 22, 1978.
108. **Dufaux, B., Hoederath, A., Heck, H., and Hollmann, W.,** Serum Testosterone Levels during the First Hours and Days after a Prolonged Physical Exercise and the Influence of Physical Training, in Abstr. 4th Int. Symp. Biochemistry of Exercise, Brussels, June 19 to 22, 1979, 8.
109. **Remes, K., Kuoppasalmi, K., and Adlercreutz, H.,** Effect of long-term physical training on plasma testosterone, androstenedione, luteinizing hormone and sex-hormone-binding globulin capacity, *Scand. J. Clin. Lab. Invest.*, 39, 743, 1979.
110. **Peltonen, P., Marniemi, J., Hietanen, E., Vuori, I., and Ehnholm, C.,** Changes in serum lipids, lipoproteins and heparin releasable lipolytic enzymes during moderate physical training in man: a longitudinal study, *Metabolism*, 30, 518, 1981.
111. **Bonen, A., Belcastro, A. N., Ling, W., Simpson, A. A., Neil, R., and MacGrail, J. C.,** Effects of Acute and Chronic Exercise on the Circulating Concentrations of Progestrone, Estradiol, FSH and LH, in Abstr. 4th Int. Symp. Biochemistry of Exercise, Brussels, June 19 to 22, 1979, 4.
112. **Dale, E., Alexander, R., Gerlach, D., and Martin, E.,** Hormone Profiles of Amenorrheic Distance Runners, in Abstr. 4th Int. Symp. Biochemistry of Exercise, Brussels, June 19 to 22, 1979, 7.
113. **Keizer, A., Poortman, J., and Bunnik, S. J.,** The Metabolic Clearance Rate of Oestradiol during Bicycle Ergometer Work with Special Reference to the Adaptation to Training, in Abstr. 4th Int. Symp. Biochemistry of Exercise, Brussels, June 19 to 22, 1979, 12.

Chapter 11

EXERCISE AND LIPOLYTIC ENZYMES

Jukka Marniemi and Eino Hietanen

TABLE OF CONTENTS

I. INTRODUCTION

In the evaluation of the mechanisms of how exercise and training modify plasma lipoproteins, the lipolytic enzymes play a key role. The association of lipoprotein lipase (LPL) with the elevation of the HDL lipoprotein fraction and simultaneously with the decrease of VLDL triglycerides has been clarified.[1–3] Although this relationship is far from solved, recent data strongly support that thc LPL enzyme is one of the rate limiting factors in the regulation of plasma lipoprotein metabolism.[4] A positive correlation has been found between plasma HDL cholesterol and postheparin plasma or adipose tissue LPL enzyme in numerous studies, both from healthy sedentary or trained persons and from those having numerous metabolic disorders.[1–3,5] It has been suggested that the lipolytic catabolism of VLDL is associated with the production of HDL precursors.[4]

The initial step in the catabolism of triglycerides is their hydrolysis by the LPL enzyme on the luminal surface of the endothelial cells of the capillaries of adipose tissue, muscles, and lungs.[4,6–11] As muscles and adipose tissue constitute major repositories of the LPL, these tissues also channel triglyceride fatty acids for oxidation or for storage. Although adipose tissue and muscle LPL have many common characteristics[12] they differ from each other in terms of their substrate affinity.[13] The Km-values for the adipose tissue LPL are about tenfold in comparison with the muscle tissue value.[13] This is well in accordance with the behavior of these tissue enzymes in response to fasting; when serum triglyceride level is low, less triglycerides will be hydrolyzed to be stored in the adipose tissue and more is directed to the muscle to be used as the energy source when hydrolyzed to fatty acids (Figure 1).[14] These types of mechanisms might also be possible in the regulation of the LPL activity during the long-term exercise when more triglycerides are used as muscle energy. Circulating as well as muscle tissue triglycerides would be hydrolyzed in the muscle and released free fatty acids would be used as energy for muscle work while simultaneously HDL would be produced in the circulation (Figure 1).

In the preceding chapters exercise and training were found to have a marked influence on plasma lipoproteins. This association has been vigorously studied both in elite athletes and in persons having moderate training like jogging and other aerobic sports.[3,15–25] Although other factors also regulate serum lipoproteins, training is one of the most physiological.[1] It has been strongly suggested that elevated HDL cholesterol levels might have protective effects to prevent the development of coronary heart disease which has increased the interest in influencing lipoproteins by physiological means.[26–34] The association between physical training and plasma HDL cholesterol is not always a direct one. In some cross-sectional studies a correlation has been found between the physical fitness and serum HDL levels.[19,20] However, in many longitudinal studies, no direct correlation has been found[3,8,35] which suggests that changes in physical fitness are not directly seen in changes in HDL cholesterol. It is more evident that physical activity is related to changes in HDL cholesterol.

II. LIPOPROTEIN LIPASE AND HORMONE-SENSITIVE LIPASE

Despite the numerous studies on the association between physical training and plasma lipoproteins, quite a few studies exist on the mechanisms of how training and exercise might mediate these changes.[1,3,35,39] Due to the link between triglyceride catabolism and formation of HDL fraction mediated by the LPL enzyme the response of this enzyme to exercise has been studied.[35,40] In experimental studies the effects of exercise on both muscle and adipose tissue lipolytic activities have been studied (Table 1).[9,10,41–43] In these studies the acute 60 to 180 min exercise increases skeletal muscle LPL activity,

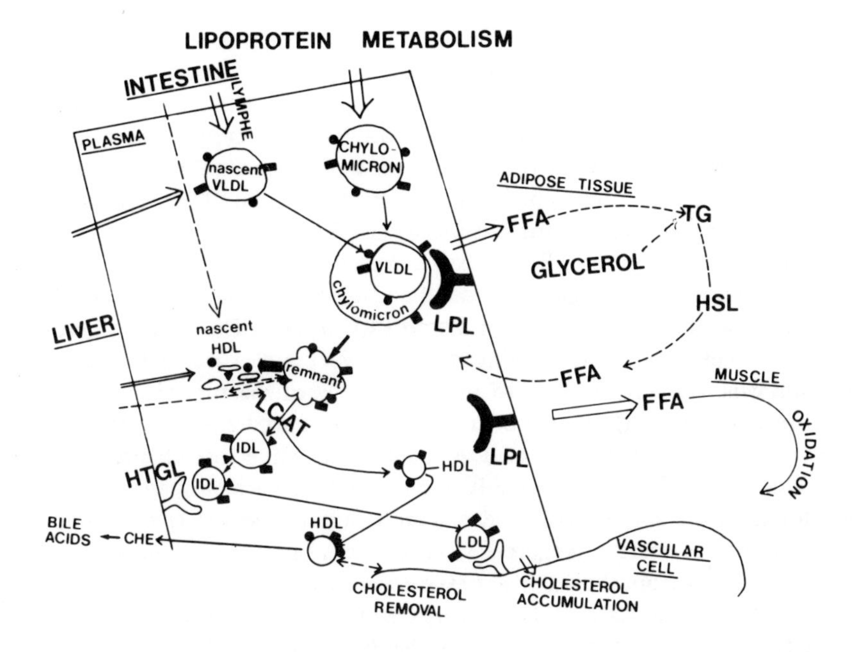

FIGURE 1. Lipoprotein metabolism in the circulation and the formation and catabolism of various lipoproteins. Abbreviations: LPL = lipoprotein lipase; LCAT = lecithin cholesterol acyltransferase; HSL = hormone sensitive lipase; HTGL = hepatic triglyceride lipase; CHE = cholesterol esters, and apoprotiens are shown as follows: ● = apo A; ▬ = apo B; ■ = apo C, and ▲ = apo E. (Modified from Leiss et al.[14]).

Table 1
THE EFFECT OF TRAINING ON LIPOPROTEIN LIPASE ACTIVITY IN ANIMALS[9,10,41–43]

	Training		% Change in Lipolytic Activity		
			Adipose tissue		Skeletal muscle
Species	Period	Time	Hormone-sensitive lipase	Lipoprotein lipase	Lipoprotein lipase
Rat	7—12 weeks	120 min	+111, +61	+4	+17
Rat	12 weeks	120			+343, +69
Rat	Single spurt	90		−34	+57[a]
Rat	Single spurt	60		−40	+69,[a] +11
Dog	Single spurt	180			+57
Rat	19—20 weeks	Progressively increasing			+34,[a] +87, +18

[a]Heart.

and markedly increased activity has been found in the cardiac LPL activity.[10,43,44] The adipose tissue LPL activity has decreased (Table 1). Kozlowski et al.[43] studied the LPL activity in skeletal muscles (quadriceps femoris) during the physical exercise in dogs. They studied the activity during a 3-hr treadmill exercise and during a 2-hr postexercise period. During the first hour the activity increased, reaching a plateau, and decreased after the exercise back to the original level in 1 hr. A 3-month regular training in rats increased skeletal muscle LPL activity while no changes were found in the cardiac LPL activity.[41]

The muscle LPL activity varies depending on the muscle type concerned. In soleus muscle the LPL activity is over twofold as compared to tibialis,[45] and the response to exercise in soleus is much more pronounced than in tibialis, at least in rats.[11,45] Prolonged exercise increases blood free fatty acids as an indication of the activation of the hormone sensitive lipase in the adipose tissue. Askew et al.[41] studied the effect of physical exercise and training to exhaustion on the adipose tissue epinephrine sensitive lipase and heart, skeletal muscle, and adipose tissue LPL in rats. The training periods were either 12 or 7 weeks, rats were running 5 days a week at 1.1 mi/hr for 120 min with a 30 sec sprint at 1.8 mi/hr every 15 min at the end of the first training group, and at the end of experiment II rats were running without sprints. The adipose tissue epinephrine stimulated lipolysis increased about twofold while less changes were found in the LPL activity (Table 1). Possibly regular exercise training increases hormone sensitive lipase in the adipose tissue also mobilizing adipose tissue fatty acids for the use in the muscles as energy. Apparently on the basis of experimental studies the metabolic flow of fats is from the adipose tissue to muscles in acute exercise favoring the use of mobilized fatty acids as muscle energy sources (Figure 1).

In man, first studies on the effect of exercise on plasma lipids were from studies in which postalimentary lipemia was found to disappear faster in persons with exercise than without.[46,47] Later the effects of ergometer training as well as out of laboratory exercise have been studied (Table 2).[38,48,49] Lithell et al.[49] have found that a 85 km skiing performance increased the skeletal muscle LPL activity by 220%. They studied the LPL activity in seven men before and after 85 km skiing race. The LPL activity increased during the race and the triglyceride droplets decreased in slow-twitch fibers during the race. The best trained subjects had the largest triglyceride droplets before the race and their droplets decreased most. These subjects had the least increase in the LPL activity while those with a low condition had the highest LPL increase. However, ergometer exercise did not increase the muscle LPL activity.[48] Holm et al.[48] did not find any difference in the LPL activity in skeletal muscle after a short-term exercise. In a study by Taskinen et al.[39] a 20 km run doubled the skeletal muscle LPL activity and also increased adipose tissue LPL activity (Table 2). No marked changes in serum lipoproteins were found due to the acute exercise (Table 2).

In cross-sectional human studies on the effect of a long-term regular physical training, marked changes in plasma lipoproteins have been found, characterized by increased HDL cholesterol levels (Table 3).[3,19,25,50,51] A uniform increase in adipose tissue LPL activity is present simultaneously (Table 3). Nikkilä et al.[3] found increased LPL levels in male long distance runners (6.10 ± 1.70 μmol FFA/hr/g in adipose tissue and 1.46 ± 0.14 in muscle) in comparison with sprinters (2.37 ± 0.30; 0.82 ± 0.14) and controls (2.22 ± 0.29; 0.85 ± 0.17). In female long-distance runners, no significant changes were present in the adipose tissue (11.4 ± 2.2; 7.94 ± 1.02), but in muscle, increased activity was found in long distance runners (1.39 ± 0.10; 0.90 ± 0.09). Also in the whole body adipose tissue and muscle LPL activities were higher in male long-distance runners than in controls, but in females only muscle LPL activity was higher. In this study the adipose tissue LPL activity correlated well with HDL cholesterol in males ($r = 0.72$) but not in females. The same was true in males for the total adipose tissue LPL activity. The skeletal muscle LPL activity correlated negatively with VLDL triglycerides ($r = -0.42$). No correlations were found between adipose tissue LPL vs. VLDL triglycerides in males and in females no correlations were present.

In a cross-sectional study by Marniemi et al.,[19] a marked increase was present in the

Table 2
RESPONSE OF LIPOPROTEIN LIPASE ACTIVITY TO ACUTE EXERCISE[38,48,49]

			Lipoprotein lipase activity (% change)		Blood lipids and lipoproteins (% change)				
							Lipoproteins		
Sex	Type of exercise	Length/ duration	Skeletal muscle	Adipose tissue	Lipids triglycerides	Cholesterol	VLDL	LDL	HDL
Male	Skiing	85 km	+220						
Male/female	Ergometer	60 min	−15	+44					
Male	Running	20 km	+112	+20	+1	+3	±0	+3	+3
	Ergometer	60 min			−16				

adipose tissue LPL activity due to physical training while there was a simultaneous decrease in the hepatic lipase activity. A similar decrease was found by Krauss et al.[50] who found a 36% decrease in the hepatic lipase while they found a 39% increase in the postheparin plasma LPL activity (Table 3).

Quite a few longitudinal studies have been made on the influence of exercise training on the enzymes regulating plasma lipoprotein metabolism despite a large amount of studies on the effect of training on plasma lipoproteins. Still it is of importance to know not only what is changed but with which mechanism the changes in serum lipoproteins take place due to training. Erkelens et al.[36] studied the effect of a moderate or low degree of exercise on plasma lipoproteins and on the postheparin lipolytic activity in a group of patients with coronary heart disease or infarction. They did not find any difference in the postheparin plasma lipolytic activity despite increase in the HDL cholesterol concentration. However, the variation in the test subjects might have hampered the possible changes. In our recent studies,[35,37] we have followed the activities of the postheparin plasma LPL and hepatic lipase as well as adipose tissue LPL along with changes in plasma lipoproteins in healthy training young and middle-aged males. These persons had a training program composed of jogging, skiing, swimming, and cycling 3 to 5 times a week, 30 to 60 min at a time.[35] The physical fitness was tested before and after the 15 weeks training program by an ergometer exercise test. When the physical fitness was plotted vs. the energy expediture used in training, there was a correlation, although not very marked, suggesting the improvement in physical condition due to exercise training; this exercise level has also been found in other studies to be sufficient.[52] The HDL cholesterol increased significantly although not very highly in magnitude due to the low level of exercise (Table 3). When the LPL activity was followed in the postheparin plasma it had already increased in the beginning of the test period and remained significantly elevated during the training period (Figure 2). The hepatic lipase did not change due to the training program (Figure 2). In this study the postheparin plasma LPL increase was associated with the decrease in the total triglyceride concentration. Moreover, there was a negative correlation between HDL cholesterol and postheparin plasma hepatic lipase both before and after the training period.[35] The relative change in HDL cholesterol also correlated positively with the change in the postheparin plasma LPL activity. This study also suggests, in addition to a positive significance of the LPL activity in the formation of HDL fraction, a negative connection between HDL fraction and hepatic lipase activity. Hepatic lipase might catalyze the removal of cholesterol from HDL particles in the liver endothelial cells.[53–55] Thus it might be favorable to have a low hepatic lipase activity in terms of elevating plasma

Table 3
EXERCISE TRAINING AND LIPOPROTEIN LIPASE ACTIVITY[3,19,25,50,51]

Type of training	Sex	Changes in lipids and lipoproteins: Triglyceride	Cholesterol	VLDL	LDL	HDL	% Changes in lipoprotein lipase activity: Postheparin plasma	Adipose tissue	Skeletal muscle	Hepatic lipase
Cross-sectional studies										
Sprinters	M	+28	+7	+45	±0	+6		+7	−4	
Long-distance runners	M	−11	+15	−6	+9	+40		+175	+72	
Long-distance runners	F	−2	+13	−7	+7	+18		+44	+54	
Jogging, skiing	M						−1	+70		−31
Long-distance runners	M						+39			−36
Longitudinal studies										
Jogging, skiing, swimming, cycling	M	−2	−4		−7	+7	+33	+56		−4
Walking, jogging (Coronary heart disease patients)	M/F	−6	−6		−4	+12	−2			

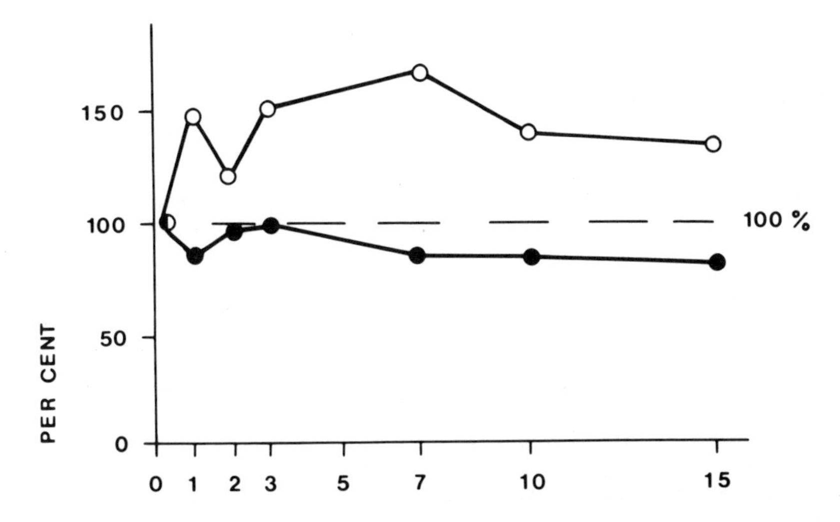

FIGURE 2. Changes in the activities of lipoprotein lipase and hepatic lipase in postheparin plasma during the course of 15 weeks conditioning period.[35] The preconditioning level is marked as 100%.

HDL cholesterol. In recent cross-sectional studies, decreased levels of hepatic lipase have been found in physically active men.[19,50]

Krauss et al.[50] studied the effect of exercise in long-distance runners on serum lipids and lipolytic activity. They found that the postheparin plasma lipolytic activity was higher in runners than in controls (5.0 ± 0.5 vs. 3.6 ± 0.3 mEq/mℓ/hr). Hepatic lipase was lower in runners (4.1 ± 0.6 vs. 6.4 ± 0.5.). Runners had higher $HDL_{2a \text{ and } b}$ than controls while no significant difference was between these groups in HDL_3. Hepatic lipolytic activity correlated in runners with HDL_3 but not in controls. Whether changes in the activity of hepatic lipase have any prognostic significance remains to be studied.

From the data in preceding articles concerning the relationship between the LPL activity and exercise training, it is to be concluded that this might, together with changes in hormone sensitive lipase, and possibly hepatic lipase, be the mechanisms of how training increases HDL cholesterol levels. However, even this is probably not a direct route of influence but other factors might be involved. Exercise training has both acute and long-term effects on many of the hormones participating in the regulation of lipolytic enzymes. Although no direct cause and consequence relationship has been proved in man in terms of hormonal mechanism in the regulation of lipolytic enzyme activities and consequently in the plasma lipoproteins, the experimental and indirect human evidence support this.

In concert with change in the physical fitness due to exercise training, physical activity might also alter personal lifestyle and dietary habits. How to exclude these factors from altering parameters supposed to be changed due to the physical fitness remains difficult to analyze. However, recent data[17,56] showed that dietary habits or intentional changes in the diet could not explain changes in plasma lipoproteins found between persons exercising by running and those of sedentary lifestyle. Also in our study,[35] dietary factors could not explain the data on the response of the LPL and hepatic lipase, but the changes of these enzymes were preferably due to the increase in physical activity.

REFERENCES

1. **Nikkilä, E. A.,** Metabolic and endocrine control of plasma high density lipoprotein concentration, Relation to catabolism of triglyceride-rich lipoproteins, in *High Density Lipoproteins and Atherosclerosis,* Gotto, A. M., Jr., Miller, N. E., and Oliver, M. F., Eds., Elsevier/North-Holland, Amsterdam, 1978, 177.
2. **Nikkilä, E. A., Taskinen, M.-R., and Kekki, M.,** Relation of plasma high-density lipoprotein cholesterol to lipoprotein lipase activity in adipose tissue and skeletal muscle of man, *Atherosclerosis,* 29, 497, 1978.
3. **Nikkilä, E. A., Taskinen, M.-R., Rehunen, S., and Härkönen, M.,** Lipoprotein lipase activity in adipose tissue and skeletal muscle of runners: relation to serum lipoproteins, *Metabolism,* 27, 1661, 1978.
4. **Tall, A. R. and Small, D. M.,** Plasma high density lipoproteins, *N. Engl. J. Med.,* 299, 1232, 1978.
5. **Nikkilä, E. A. and Hormila, P.,** Serum lipids and lipoproteins in insulin-treated diabetes. Demonstration of increased high density lipoprotein concentrations, *Diabetes,* 27, 1078, 1978.
6. **Hartiala, J., Viikari, J., Hietanen, E., Toivonen, H., and Uotila, P.,** Cigarette smoke affects lipolytic activity in isolated rat lungs, *Lipids,* 15, 539, 1980.
7. **Hietanen, E. and Greenwood, M. R. C.,** A comparison of lipoprotein lipase activity and adipocyte differentiation in growing male rats, *J. Lipid Res.,* 18, 480, 1977.
8. **Hietanen, E. and Hartiala, J.,** Developmental pattern of pulmonary lipoprotein lipase in growing rats, *Biol. Neonate,* 36, 85, 1979.
9. **Nikkilä, E. A., Torsti, P., and Penttilä, O.,** The effect of exercise on lipoprotein lipase activity of rat heart, adipose tissue and skeletal muscle, *Metabolism,* 12, 863, 1963.
10. **Nikkilä, E. A., Torsti, P., and Penttilä, O.,** Effects of fasting, exercise and reserpine on catecholamine content and lipoprotein lipase activity of rat heart and adipose tissues, *Life Sci.,* 4, 27, 1965.
11. **Rauramaa, R., Kuusela, P., and Hietanen, E.,** Adipose, muscle and lung tissue lipoprotein lipase activities of young streptozotocin treated rats, *Horm. Metab. Res.,* 12, 591, 1980.
12. **Brady, M. and Higgins, J. A.,** The properties of the lipoprotein lipases of rat heart, lung, and adipose tissue, *Biochim. Biophys. Acta,* 137, 140, 1967.
13. **Fielding, C. J.,** Lipoprotein lipase: evidence for high- and low affinity enzyme sites, *Biochemistry,* 15, 879, 1976.
14. **Leiss, O., Murawski, U., and Egge, H.,** Lecithin: cholesterol acyltransferase activity in relation to lipoprotein concentration and lipid composition, *Scand. J. Clin. Lab. Invest.,* 38(Suppl. 150), 77, 1978.
15. **Farrell, P. A. and Barboriak, J.,** The time course of alterations in plasma lipid and lipoprotein concentrations during eight weeks of endurance training, *Atherosclerosis,* 37, 231, 1980.
16. **Hartung, G. H., Foreyt, J. P., Mitchell, R. E., Vlasek, J., and Gotto, A. M., Jr.,** Relationship of diet and HDL cholesterol in sedentary and active middle-aged men, *Circulation,* 58(Suppl. 2), 204, 1978.
17. **Hartung, G. H., Foreyt, J. P., Mitchell, R. E., Vlasek, I., and Gotto, A. M., Jr.,** Relation of diet to high-density-lipoprotein cholesterol in middle-aged marathon runners, joggers, and inactive men, *N. Engl. J. Med.,* 302, 357, 1980.
18. **Huttunen, J. K., Länsimies, E., Voutilainen, E., Ehnholm, Chr., Hietanen, E., Penttilä, I., Siitonen, O., and Rauramaa, R.,** Effect of moderate physical exercise on serum lipoproteins. A controlled clinical trial with special reference to serum high density lipoproteins, *Circulation,* 60, 1220, 1979.
19. **Marniemi, J., Peltonen, P., Vuori, I., and Hietanen, E.,** Lipoprotein lipase of human postheparin plasma and adipose tissue in relation to physical training, *Acta Physiol. Scand.,* 110, 131, 1980.
20. **Miller, N. E., Rao, S., Lewis, B., Björsvik, G., Myhre, K., and Mjös, O. D.,** High-density lipoprotein and physical activity, *Lancet,* 1, 111, 1979.
21. **Schwane, J. A. and Cundiff, D. E.,** Relationships among cardiorespiratory fitness, regular physical activity and plasma lipids in young adults, *Metabolism,* 28, 771, 1979.
22. **Sutherland, W. H. F. and Woodhouse, S. P.,** Physical activity and plasma lipoprotein lipid concentrations in man, *Atherosclerosis,* 37, 285, 1980.
23. **Vodak, P. A., Wood, P. D., Haskell, W. L., and Williams, P. T.,** HDL-Cholesterol and other plasma lipid and lipoprotein concentrations in middle-aged male and female tennis players, *Metabolism,* 29, 745, 1980.
24. **Wood, P. D. and Haskell, W. L.,** The effect of exercise on plasma high density lipoproteins, *Lipids,* 14, 417, 1979.
25. **Wood, P. D., Haskell, W., Klein, H., Lewis, S., Stern, M. P., and Farquhar, J. W.,** The distribution of plasma lipoproteins in middle-aged male runners, *Metabolism,* 25, 1249, 1976.

26. **Barboriak, J. J., Anderson, A. J., Rimm, A. A., and King, J. F.,** High density lipoprotein cholesterol and coronary artery occlusion, *Metabolism,* 28, 735, 1979.
27. **Brunner, D., Weisbort, J., Loeb, K., Schwartz, S., Altman, S., Bearman, J. E., and Levin, S.,** Serum cholesterol and high density lipoprotein cholesterol in coronary patients and healthy persons, *Atherosclerosis,* 33, 9, 1979.
28. **Carneiro, R. C., Lion, M. F., and Melo, E.,** Alteraqoes lipidicas e insuficiencia coronariana. Relaqao entre dados laboratoriais e cinecoronariografia, *Arq. Bras. Cardiol.,* 31, 127, 1978.
29. **Castelli, W. P., Doyle, J. T., Gordon, T., Hames, C. G., Hjortland, M. C., Hulley, S. B., Kagan, A., and Zukel, W. J.,** HDL cholesterol and other lipids in coronary heart disease. The cooperative lipoprotein phenotyping study, *Circulation,* 55, 767, 1977.
30. **Gordon, T., Castelli, W. P., Hjortland, M. C., Kannel, W. B., and Dawber, T. R.,** High density lipoprotein as a protective factor against coronary heart disease, The Framingham study, *Am. J. Med.,* 62, 707, 1977.
31. **Kaukola, S., Manninen, V., and Halonen, P. I.,** Serum lipids with special reference to HDL cholesterol and triglycerides in young male survivors of acute myocardial infarction, *Acta Med. Scand.,* 208, 41, 1980.
32. **Mariotti, S., Verdecclua, A., Capocaccia, R., Conti, S., Farchi, G., and Menotti, A.,** Identificazione di soggetti ad alto rischio coronarico nel progetto romano di prevenzione della cardiopatia coronarica, *G. Ital. Cardiol.,* 7, 141, 1977.
33. **Miller, N. E.,** The evidence for the antiatherogenicity of high density lipoprotein in man, *Lipids,* 13, 914, 1978.
34. **Schonfeld, G.,** Lipoproteins in atherogenesis, *Artery,* 5, 305, 1979.
35. **Peltonen, P., Marniemi, J., Hietanen, E., Vuori, I., and Ehnholm, C.,** Changes in serum lipids, lipoproteins and heparin releasable lipolytic enzymes during moderate physical training in man. A longitudinal study, *Metabolism,* 30, 518, 1981.
36. **Erkelens, D. W., Albers, J. J., Hazzard, W. R., Frederick, R. C., and Bierman, E. L.,** High-density lipoprotein-cholesterol in survivors of myocardial infarction, *J.A.M.A.,* 242, 2185, 1979.
37. **Hietanen, E., Marniemi, J., Peltonen, P., Suistomaa, U., Viljanen, T., and Vuori, I.,** HDL Cholesterol changes in moderate physical training mediated by lipoprotein lipase, hepatic lipase and lecithin-cholesterol acyltransferase, *Publ. Univ. Kuopio/Med.,* Nr 1, 81, 1980.
38. **Lithell, H., Örlander, J., Schéle, R., Sjödin, B., and Karlsson, J.,** Changes in lipoprotein-lipase activity and lipid stores in human skeletal muscle with prolonged heavy exercise, *Acta Physiol. Scand.,* 107, 257, 1979.
39. **Taskinen, M.-R., Nikkilä, E. A., Rehunen, S., and Gordin, A.,** Effect of acute vigorous exercise on lipoprotein lipase activity of adipose tissue and skeletal muscle in physically active men, *Artery,* 6, 471, 1980.
40. **Nikkilä, E. A. and Taskinen, M.-R.,** Relation between H.D.L.-cholesterol levels and triglyceride metabolism, *Lancet,* 2, 892, 1978.
41. **Askew, E. W., Dohm. G. L., Huston, R. L., Sneed, T. W., and Dowdy, R. A.,** Response of rat tissue lipases to physical training and exercise, *Proc. Soc. Exp. Biol. Med.,* 141, 123, 1972.
42. **Borensztajn, J., Rone, M. S., Babirak, S. P., McGarr, J. A., and Oscai, L. B.,** Effect of exercise on lipoprotein lipase activity in rat heart and skeletal muscle, *Am. J. Physiol.,* 229, 394, 1975.
43. **Kozlowski, S., Budohoski, L., Pohoska, E., and Nazar, K.,** Lipoprotein lipase activity in the skeletal muscle during physical exercise in dogs, *Pflügers Arch.,* 382, 105, 1979.
44. **Fukuda, N., Ide, T., Kida, Y., Takamine, K., and Sugano, M.,** Effects of exercise on plasma and liver lipids of rats. IV. Effects of exercise on hepatic cholesterogenesis and fecal steroid excretion in rats, *Nutr. Metab.,* 23, 256, 1979.
45. **Pařízková, J.,** Lipoprotein lipase activity in heart and skeletal muscle after adaptation to different loads, in *Body Fat and Physical Fitness,* Pařízková, J., Ed., Martinus Nijhoff, The Hague, 1977, 90.
46. **Cohen, H. and Goldberg, C.,** Effect of physical exercise on alimentary lipaemia, *Br. Med. J.,* 2, 509, 1960.
47. **Nikkilä, E. A. and Konttinen, A.,** Effect of physical activity on postprandial levels of fats in serum, *Lancet,* 1, 1151, 1962.
48. **Holm, G., Jacobsson, B., Holm, J., Björntorp, P., and Smith, U.,** Effects of submaximal physical exercise on adipose tissue metabolism in man, *Int. J. Obesity,* 1, 249, 1977.
49. **Lithell, H., Hellsing, K., Lundqvist, G., and Malmberg, P.,** Lipoprotein-lipase activity of human skeletal-muscle and adipose tissue after intensive physical exercise, *Acta Physiol. Scand.,* 105, 312, 1979.
50. **Krauss, R. M., Wood, P. D., Giotas, C., Waterman, D., and Lindgren, F. T.,** Heparin-released plasma lipase activities and lipoprotein levels in distance runners, *Circulation,* 60(Suppl. 2) 73, 1979.

51. **Costill, D. L., Fink, W. J., Getchell, L. H., Ivy, J. L., and Witzmann, F. A.,** Lipid metabolism in skeletal muscle of endurance-trained males and females, *J. Appl. Physiol.,* 47, 787, 1979.
52. **Wilmore, J. H., Davis, J. A., O'Brien, R. S., Vodak, P. A., Walder, G. R., and Amsterdam, E. A.,** Physiological alterations consequent to 20-week conditioning programs of bicycling, tennis, and jogging, *Med. Sci. Sports Exercise,* 12, 1, 1980.
53. **Jansen, H., van Tol, A., and Hülsman, W. C.,** On the metabolic function of heparin-releasable liver lipase, *Biochem. Biophys. Res. Commun.,* 92, 53, 1980.
54. **Kinnunen, P. K. J.,** High-density lipoprotein may not be antitherogenic after all, *Lancet,* 2, 34, 1979.
55. **Kuusi, T., Kinnunen, P. K. J., and Nikkilä, E. A.,** Hepatic endothelial lipase antiserum influences rat plasma low and high density lipoproteins in vivo, *FEBS Lett.,* 104, 384, 1979.
56. **Kiens, B., Gad, P., Lithell, H., and Vessby, B.,** HDL-cholesterol, physical activity and diet in middle-aged men, *Acta Physiol. Scand. Suppl.,* 108(Abstr.), 34, 1980.

Chapter 12

RESPONSE OF SERUM LECITHIN CHOLESTEROL ACYLTRANSFERASE ACTIVITY TO EXERCISE TRAINING

Jukka Marniemi and Eino Hietanen

TABLE OF CONTENTS

I. INTRODUCTION

Lecithin cholesterol acyltransferase (LCAT) is one of the key enzymes in lipoprotein metabolism. It transfers fatty acids from 2-position of lecithin to unesterified cholesterol in nascent high density lipoprotein (HDL) fraction.[1,2] In addition to its principal substrate, HDL, LCAT also indirectly affects the composition of other lipoproteins by promoting the nonenzymatic transfer of cholesteryl esters from HDL to very low density (VLDL) and low density (LDL) lipoproteins.[3] Apolipoprotein distribution and probably some other aspects of lipoprotein metabolism are also influenced by LCAT reaction.[3]

Despite its regulatory role in HDL metabolism, quite little attention has been paid to LCAT in studies concerning the biochemical mechanisms of increased HDL by physical exercise. Significantly enhanced LCAT activities (about 50%) have been recently reported in rats during a 2-month swimming program.[4] In man one preliminary report with four test subjects has been published on the effect of physical training on plasma LCAT activity.[5] In the study by Lopez et al.[5] a 30 min exercise four times a week resulted in an increase in the LCAT activity from 0.38 μmol/mℓ/hr to 0.52 μmol/mℓ/hr. In our present longitudinal training study where parameters of lipid metabolism were followed, changes in LCAT activity were also measured.

II. EXERCISE AND LECITHIN CHOLESTEROL ACYLTRANSFERASE

The training group in our study consisted of healthy, normolipemic (mean age 38.5 years, range 31 to 48 years) men (N = 19) selected randomly from the screening material of a local physical fitness research center.[6] A reference group of six men was also included in the study to exclude any possible seasonal variations. Both groups had a sedentary lifestyle with a low amount of physical activity before the test period. In the beginning of the study the two groups did not differ from each other in relation to relative body weight, age, or physical fitness. The training group exercised regularly for 15 weeks at least three times a week (30 to 60 min each time) under the guidance of a physical educator.[6] The controlled training was composed of aerobic exercises, such as jogging and cross-country skiing. During the test period the reference subjects retained their normal physical activity. The participants were requested not to change their eating habits during the follow-up period which was also controlled with a questionnaire before and after the period. A pulse conducted bicycle ergometer test up to a subjective maximum was performed on the subjects of both groups before and after the test period.[7] The work load at pulse rate 150/min was used as an index of physical performance capacity.

Fasting blood samples were taken from the trainers once a week during the first training month and then once a month to the end of the test period. From the reference group the samples were taken only before and after the test period. Serum samples were stored at −20°C until assayed at the same time after the test period to avoid inter-assay error in the determinations. The LCAT enzyme is difficult to determine in a laboratory. It is known that lipoproteins activate this enzyme.[8] Thus in the presence of the lipoproteins in the serum, where the enzyme activity is to be determined, a possibility exists that, in addition to the enzyme activity, one also assays the response of apoproteins to exercise training. In order to avoid this difficulty in interpreting the data we have used delipoproteinized enzyme source instead of whole serum or plasma. The LCAT activity was determined according to Alcindor et al.[9,10] after delipoproteinization of the serum samples, with Intralipid®, dextran sulfate and calcium chloride. The substrate was prepared from pooled human serum by β-delipoproteinization with dextran sulphate and $CaCl_2$ followed by addition of tritiated cholesterol, inhibition of LCAT by heating, and

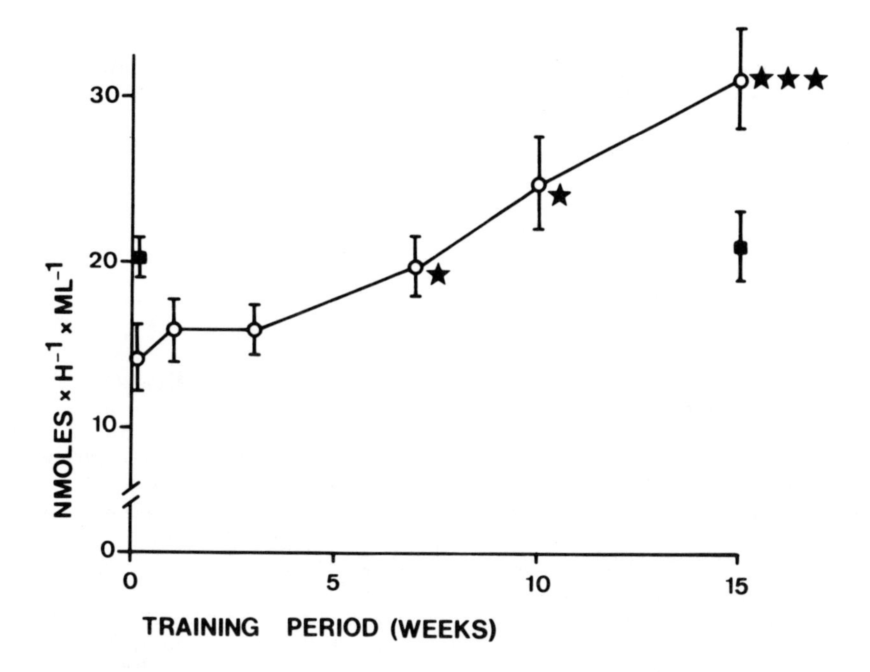

FIGURE 1. The serum LCAT activity during the 15 weeks training period expressed as nanomoles of cholesterol esterified $\times$ h^{-1} $\times$ ml^{-1}. The activities of the sedentary reference subjects (black squares) in the beginning and at the end of the follow-up are also shown. Means and their standard errors (vertical lines) are given. The significance of the differences compared to the starting level are as follows: x = $p < 0.05$, xxx = $p < 0.001$.

equilibration of radioactive cholesterol to HDL-fraction by an incubation at +5°C overnight.

The effectiveness of the training was indicated by the significant increase (14%) in physical performance capacity obtained in the exercise group as judged by work load at the heart rate 150/min, while no change occurred in the reference group.[6] HDL cholesterol, HDL/total cholesterol ratio, postheparin plasma and adipose tissue LPL activities increased significantly, while in triglycerides, total cholesterol, and postheparin plasma HL a tendency, although not statistically significant, to lowered levels was obtained.[6] After 7 weeks training the increment in the LCAT activity was already significant and after the 15 week program the mean activity was over twofold compared to the starting level (Figure 1). During the same time the enzyme activity did not change in the sedentary reference group. Simko and Kelley[4,11] have shown that in rats exercise enhances the LCAT activity both on a low cholesterol diet and on a diet high in cholesterol and sucrose. These data suggest that at least in rats the physical exercise as such rather than the increased food intake is the basic modifier of the LCAT activity. Whether this is also true in man remains to be solved.

How this enhancement of LCAT activity by physical exercise is connected to the higher HDL levels in physically active subjects can only be speculated at present. In some recent studies no positive correlation between the LCAT activity in vitro and the plasma HDL cholesterol has been found.[12,13] Neither has any connection of the LCAT activity with the myocardial infarction been found.[14] Our present data indicated that during a moderate aerobic training program the serum LCAT activity is significantly increased in previously sedentary subjects in accordance with the data by Lopez et al.[5]

Whether exercise training has a direct influence on the LCAT enzyme or whether the changes in the enzyme activity, as well as changes found in the lipoprotein lipase (LPL) enzyme activity, are due to the direct effects or mediated by hormones changing in response to exercise training remains to be solved. Only a little is thus far known about the regulation of the LPL activity, e.g., in comparison with the LCAT activity which is studied far more.

REFERENCES

1. **Glomset, J. A.,** The mechanisms of plasma cholesterol esterification reaction: plasma fatty acid transferase, *Biochim. Biophys. Acta.*, 65, 128, 1962.
2. **Glomset, J. A.,** The plasma lecithin: cholesterol acyltransferase reaction, *J. Lipid. Res.*, 9, 155, 1968.
3. **Glomset, J. A.,** Lecithin cholesterol acyltransferase, in *The Biochemistry of Atherosclerosis,* Scanu, A. M., Wissler, R. W., and Getz, G. S., Eds., Marcel Dekker, New York, 1979, 247.
4. **Simko, V. and Kelley, R. E.,** Effect of chronic intermittent exercise on biliary lipids, plasma lecithin cholesterol acyltransferase, and red blood cell lipids in rats, *Am. J. Clin. Nutr.*, 32, 1376, 1979.
5. **Lopez-S, A., Vial, R., Balart, L., and Arroyave, G.,** Effect of exercise and physical fitness on serum lipids and lipoproteins, *Atherosclerosis,* 20, 1, 1974.
6. **Peltonen, P., Marniemi, J., Vuori, I., and Hietanen, E.,** Changes in serum lipids, lipoproteins and heparin releasable lipolytic enzymes during moderate physical training in man. A longitudinal study, *Metabolism.*, 30, 518, 1981.
7. **Arstila, M.,** Pulse-conducted triangular exercise ECG-test, *Acta Med. Scand. Suppl.*, 529, 1, 1972.
8. **Leiss, O., Murawski, U., and Egge, H.,** Lecithin: cholesterol acyltransferase activity in relation to lipoprotein concentration and lipid composition, *Scand. J. Clin. Lab. Invest.*, 38(Suppl. 150), 77, 1979.
9. **Alcindor, L. G., Melin, B., Benhamou, G., and Piot, M. C.,** Dosage de la lecithine-cholesterol acyltransferase. Interet de la precipitation des β-lipoproteins par le sulfate de dextrane, *Clin. Chim. Acta.*, 81, 177, 1977.
10. **Alcindor, L. G., Dusser, A., Piot, M. C., Infante, R., and Polonowski, J.,** A rapid method for lecithin: cholesterol acyltransferase estimation in human serum, *Scand. J. Clin. Lab. Invest.*, 38(Suppl. 150), 12, 1978.
11. **Simko, V. and Kelley, R. E.,** Physical exercise modifies the effect of high cholesterol-sucrose feeding in the rat, *Eur. J. Appl. Physiol.*, 40, 145, 1979.
12. **Soloff, L. A. and Varma, K. G.,** Relationship between high density lipoproteins and the rate of in vitro serum cholesterol esterification, *Scand. J. Clin. Lab. Invest.*, 38(Suppl. 150), 72, 1978.
13. **Stokke, K. T. and Enger, S. C.,** Plasma HDL cholesterol and the *in vitro* esterification of cholesterol, *Scand. J. Clin. Lab. Invest.*, 39, 597, 1979.
14. **Wiklund, O. and Gustafson, A.,** Lecithin: cholesterol acyltransferase (LCAT) and fatty acid composition of lecithin and cholesterol esters in young male myocardial infarction survivors, *Atherosclerosis,* 33, 1, 1979.

Chapter 13

RELATION BETWEEN SERUM LIPIDS AND PHYSICAL EXERCISE IN CHILDREN

Jorma Viikari, Veli Ylitalo, and Ilkka Välimäki

TABLE OF CONTENTS

I. INTRODUCTION

Physical inactivity is epidemiologically considered a risk factor of coronary artery disease[1] (CAD). Physical activity is considered a protective factor, respectively. In adults the physical activity can be divided into physical activity at work and physical activity during leisure. Both are beneficial to the plasma lipoprotein metabolism,[2–5] but only the activity during leisure is protective, whereas strenuous physical activity at work appears to be a risk factor of CAD.[6]

In this article we review current data on the relation between physical activity and the lipoprotein metabolism in children and adolescents. First the special features of the lipoprotein metabolism in children are described. Next we give a short review of the methods for estimating the physical activity of children and finally the relations between physical activity and the lipoprotein metabolism in children are discussed.

II. LIPOPROTEIN METABOLISM IN CHILDREN AND ADOLESCENTS

Cord blood serum *cholesterol* levels are universally approximately 2 mmol/ℓ.[7–10] About 40% of the serum total cholesterol (TC) is in the high density lipoprotein (HDL) fraction and about 60% in the low density lipoprotein (LDL), and the very low density lipoprotein (VLDL) fractions. The TC increases with age depending somewhat on the diet. This increase is slower in those fed on formulas than in those who are breastfed.[10–12] The TC level is about 3.5 to 4.5 mmol/ℓ at 6 months in western countries and about 4.0 to 4.5 mmol/ℓ at the age of 1 year.[13–14] During the first years the TC gradually increases to the level which is maintained essentially constant throughout childhood until puberty, i.e., about 4 to 5 mmol/ℓ.[9,14,15]

Studies of adults have demonstrated that TC levels vary greatly in different countries.[16] In a recent international study coordinated by a Dutch group,[15] it was shown that there are already great differences in the TC levels of children. This variation resembles the finding in adults: 8-year-old Finnish boys had the highest TC values in 16 countries (Table 1). It has been proposed that the *ratio of HDL/TC* would indicate the risk of CAD better than the TC.[17] Therefore, it was interesting and surprising that this ratio was essentially similar in all participating countries, about 30%.

In puberty the TC decreases a little[9,14,18] and the same happens with the HDL of boys, which results in a lower HDL/TC ratio in boys than in girls. After puberty the TC increases again and by the age of 20 it reaches the childhood level.[14] The serum *triglyceride* level is usually very low in children,[14,19] but it begins to increase somewhat earlier than the serum cholesterol (Figure 1). The *apoprotein* concentrations are very important because they activate the central enzymes of the lipoprotein metabolism. Apo C II is considered the activator of the lipoprotein lipase (LPL) and Apo A I the activator of the lecithin-cholesteryl acyl transferase, respectively.[20] Furthermore, there is increasing evidence of a positive correlation between the LPL activity and the HDL level.[21] The systematic knowledge of the apoprotein levels in children is, however, still scanty.

The knowledge of the HDL is also increasing concerning children. The term "HDL" is actually ultracentrifugal, but conventional methods[22] are time-consuming and expensive. Some faster and simpler methods have been developed,[23] using the ultracentrifugal technique with gradient. Also, quantitative lipoprotein electrophoresis techniques[24] may prove practical. Simple precipitation methods like heparin-manganese chloride, dextran sulphate, or polyethylene glycol are now widely used. In practice they give qualitatively rather similar results, but quantitatively they differ so much that it is necessary to know which method has been used in specific studies.[25]

A significant degree of tracking has been reported for cholesterol and triglycerides

Table 1
TOTAL AND HDL-CHOLESTEROL CONCENTRATION IN SERUM OF BOYS AGED 7 TO 8 YEARS IN RURAL AREAS OF DIFFERENT COUNTRIES

Country	Serum cholesterol mmol/ℓ	HDL-cholesterol mmol/ℓ	HDL-cholesterol / Serum-cholesterol
Ghana	3.13	0.92	0.30
Pakistan	3.31	0.78	0.24
Portugal	3.76	1.20	0.32
U.S.	4.35	1.37	0.32
Netherlands	4.56	1.46	0.33
Sweden	4.60	1.54	0.34
Finland	5.16	1.68	0.33

Modified version of original Table 1 from Knuiman, J. T., Hermus, J. J., and Hautvast, J. G. A. J., *Atherosclerosis*, 36, 529, 1980. With permission.

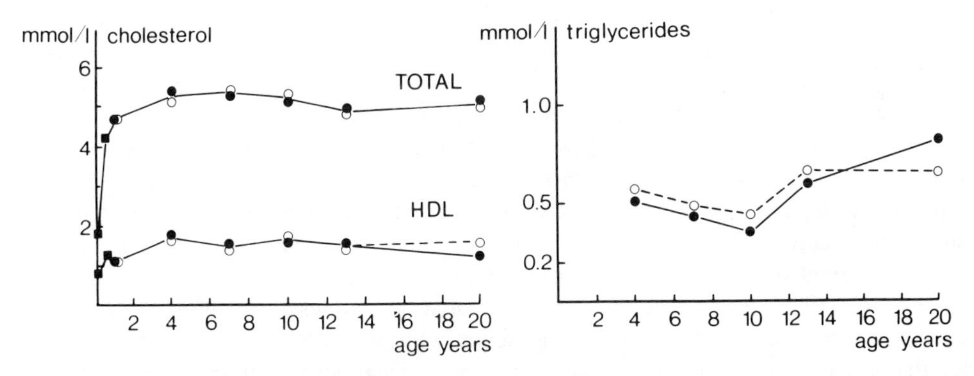

FIGURE 1. Serum lipids in Finnish children and young adults.

as well as for the HDL- and LDL-cholesterol, but only half or even less of the children initially in the top percentiles will retain persistent elevations of lipids and lipoproteins.[26,27]

III. CHILDHOOD OBESITY AND BLOOD LIPIDS

Obese children[9,28] as well as obese adults[29] tend to have higher levels of triglycerides, VLDL cholesterol, and LDL cholesterol and lower levels of HDL cholesterol than lean subjects. Finnish obese schoolchildren aged 7 to 14 years had a serum cholesterol level of 5.23 mmol/ℓ, the HDL cholesterol was 1.31 mmol/ℓ, the ratio HDL/TC was 0.25 and the serum triglycerides 0.85 mmol/ℓ.[28] A low but statistically significant correlation has already been found between the cholesterol and physical fitness in adolescence, especially in boys.[30]

IV. ESTIMATION OF PHYSICAL ACTIVITY AND PHYSICAL FITNESS IN CHILDREN

There are several ways to estimate the *habitual physical activity,* such as measurement of the oxygen consumption, recording of motion, pedometric measurements, motion picture sampling, tape recording of electrocardiogram or pneumogram, etc.[31–34] Questionnaires and time budget diaries are technically less demanding for gathering information on physical activity. They are used in population studies, but they are inferior to the former, technical methods with relation to reliability.[34] The various methods used to determine *physical fitness* or working capacity are considerably more quantitative and accurate. The degree of physical fitness can be measured by an ergometer or a treadmill. An ideal parameter to monitor would be the cardiac output, but this is fairly difficult during exercise. It has become generally accepted to measure heart rate and oxygen uptake instead, *at various submaximal levels or by a maximal test.*[35–37] The determination of the maximal oxygen uptake is feasible in children[36,37] but it is inconvenient and in a way "invasive", especially at an early school age. To avoid these inconveniences, techniques based on monitoring the heart rate only have been tried out.[35,38–40] These methods have proved appropriate for demonstrating the increase of the working capacity in relation to the increase in age and size of the child, but to discriminate between inactive and trained children maximal tests are absolutely essential. According to our experience the pulse-conducted exercise test of Arstila[41] appeared a maximal test when properly automated.[42,43] This test reasonably well discriminated children engaged in different degrees of physical activity.[44–46]

When measuring the physical activity in relation to metabolic parameters such as the lipid profile, the experimental design must be carefully selected. While the changes in blood lipids appear to be related to the everyday dosage of exertion and the lipid pattern should be compared with this, there is a clear correlation between the work capacity and the amount of training.[36,47] The main emphasis must be laid on the determination of spare time activity and/or the net effect of it, the fitness, because the activity at school is approximately the same for all school children. We think that there are three alternatives for relating the lipid pattern to the physical activity in children:

1. Large population study, where the influence of food, age, sex, physical activity, etc. is investigated by multiple discrimination analysis. In this case the physical activity is classified by means of an interview concerning the behavior of the subject
2. Study of the children before, during, and after a preselected training program. In this way each subject is his own control and the metabolic response is displayed individually as a function of the cumulative effort.[48]
3. Study, where children preclassified according to criteria based on habitual activity (trained athletes vs. inactive) are subjected to a multiparametric exercise test procedure.[44–46]

The last method is rather quantitative, but it is restricted to small test groups because of the heavy laboratory test drill. Method 2 is also applicable to small groups of subjects for the same reason. It is also susceptible to the changes caused by other factors like age, which obviously affect both the ergometric result and the lipid analysis. The first epidemiological method is the proper way for accumulating a lot of data, but the reliability and sensitivity are lower. In addition to the amount of training, the type of training selected is also critical.[49] After puberty the variance of the results in school sports tends to increase in normal children indicating a changing pattern of participation in sports.[50]

In conclusion, when the lipid pattern is related to the physical activity of the child, it is important to pay attention to the experimental design of the determination of the physical activity, the character of the training, the character of the exercise test (if applied), and the age of the subject. Although the repeat variability of the determination of the lipids and the physical activity must be small to enhance methodological accuracy, the growth, biological interindividual variation, and the multifactorial influences on the blood lipids greatly interfere with the detailed analysis of the dependence of lipid patterns on physical activity in childhood.

V. PHYSICAL ACTIVITY IN OBESE CHILDREN

In numerous studies it has been stated that obese children are physically less active as compared to children of normal weight.[51,52] Bullen et al.[31] studied the physical activity by motion picture sampling and they found that obese girls are far more inactive than the nonobese even during supervised sport periods. The same inactivity has been found also in obese infants.[53]

Obese children often have reduced motor activity and an obese child imposes a much greater load on his cardiovascular system during exercise.[54] All physical activity is thus a much greater strain for the obese child. The procedure of relating maximal oxygen uptake or working capacity to body weight penalizes the obese individual but this penalty is justified, since the energy cost of most activities is markedly influenced by the body weight.[55] The economy of work with a medium work load is substantially poorer in obese children and this is manifested by a raised heart rate and a higher oxygen consumption when performing the same work as compared to leaner children. The total performance is lower when the maximal oxygen uptake is reached.[56] A negative linear correlation exists between overweight and working capacity and physical efficiency.[57–59] In children, training increases the aerobic power and physical fitness as well as influences the body composition in such a way that the lean body mass is preserved.[36,56]

VI. RELATION BETWEEN SERUM LIPIDS AND PHYSICAL ACTIVITY

Epidemiological and other studies in adults have shown certain associations between the degree of physical activity and the serum lipids. The serum cholesterol (TC) and triglyceride levels tend to be lower in those engaged with more physical activity[3,4,17,60] although this is not a constant finding.[61] This is due to the fact that the LDL tends to decrease and the HDL, respectively, to increase with increasing physical activity. Therefore the TC may be even higher in those with a high degree of physical activity. Huttunen et al.[2] have shown that a change in physical habits is beneficial to the individual although the degree of physical activity must be high[3] before the HDL level becomes clearly elevated. Interestingly, the sera from subjects involved in a high degree of physical activity, seem to be less "atherogenic" in cell cultures.[62] If these findings can also be ascertained in children, it means that especially in the western countries we should try to increase the level of habitual physical activity in children.

VII. AMOUNT OF PHYSICAL ACTIVITY AND SERUM LIPIDS

There are rather few population studies where it has been possible to show correlations between physical activity or physical fitness and serum lipid fractions. This may be due to the fact that in random samples there are rather few individuals involved in an extremely high or a very low level of physical activity, most having adapted a reasonably high level. Lee[63] studied 118 teenagers (12 to 19 years of age), both boys and

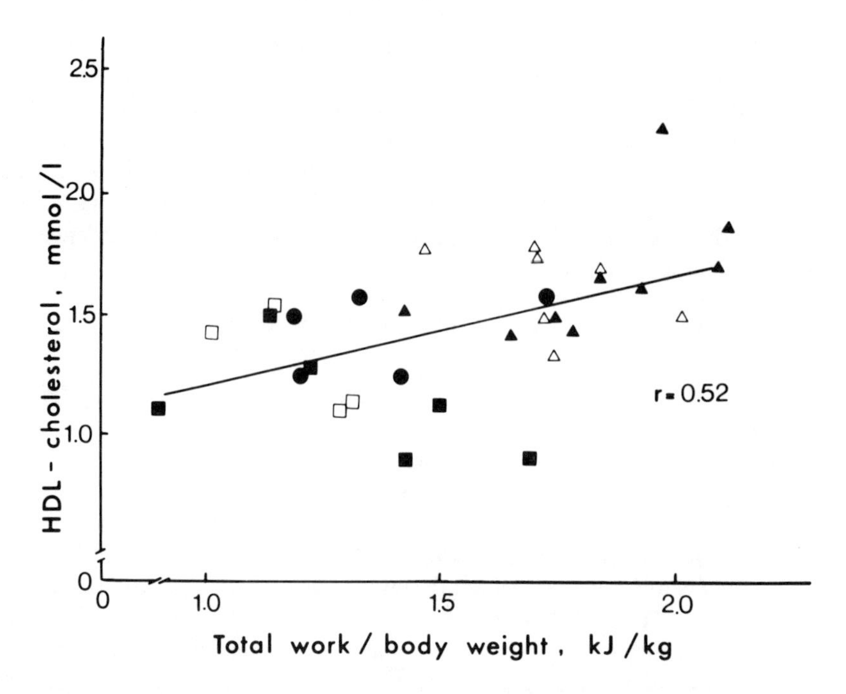

FIGURE 2. Correlation between HDL-cholesterol and total work/body weight in 11 to 13 year-old children. (r = 0.52, $p<0.01$) (▲) trained boys, (■) normally active boys, (●) physically inactive boys, (△) trained girls, (□) normally active girls.

girls. The serum cholesterol and beta lipoprotein (LDL) cholesterol were negatively associated with the physical activity index for the boys, but not for the girls. In another study 152 college students (mean age 19 years) were examined.[64] The cardiorespiratory fitness was estimated by a treadmill test. The serum triglyceride level correlated negatively with the duration of the treadmill test in the males but not in the females, whereas the HDL cholesterol correlated positively with the treadmill test in both males and females.

Gilliam et al.[65] made a thorough study of 47 active boys and girls (aged 7 to 12 years). They could not find any significant association between the maximal oxygen consumption (VO_2 max) and the serum lipids. They did not, however, in this study measure the HDL cholesterol, which could have been of value, because the negative correlation between serum triglycerides and the VO_2 max was rather high (possibly significant at level 0.05). The cholesterol and triglyceride values were also unrelated to the VO_2 max in another study,[66] which comprised 329 subjects aged 10 to 19 years. Välimäki et al.[46] selected 37 school-children on a different level of training. The TC was about 0.3 mmol/ℓ higher in the trained than in the nontrained children. This difference was obviously due to a higher HDL cholesterol level in the trained children, and a positive correlation between the HDL cholesterol concentration and the ergometric exercise tolerance could be demonstrated (Figure 2). In younger age-group (7 to 8 years) the same could not be demonstrated.[45] Rönnemaa et al.[67] examined a material of young athletes (males, mean age 17 years). The athletes were found to have slightly lower TC values than the controls. This was due to a lower LDL cholesterol; the HDL cholesterol levels were similar. When the sera of the athletes and controls were compared under cell culture conditions, no difference could be seen in the "atherogenecity", which suggests that the advantageous effects of sera from adult runners on smooth

Table 2
EFFECT OF 16-WEEK WALKING PROGRAM ON PLASMA LIPID LEVELS IN OBESE YOUNG MEN

Lipid fraction	Before	After
Total cholesterol	4.52	4.37 mmol/ℓ
HDL-cholesterol	0.83	0.96 mmol/ℓ
HDL/total	0.18	0.22
Triglycerides	1.86	1.63 mmol/ℓ

Adapted from Leon, A. S., Conrad, J., Hunninghake, D. B., and Serfass, R., *Am. J. Clin. Nutr.*, 33, 1776, 1979.

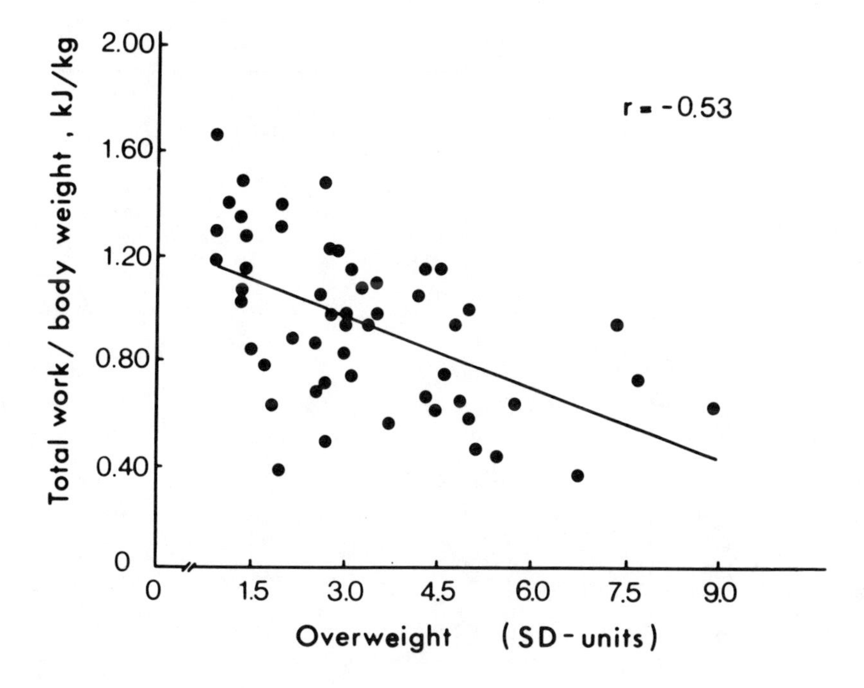

FIGURE 3. Correlation between total work/body weight and overweight ($r = -0.53$, $p<0.001$). Overweight is given in units of standard deviation. (From Ylitalo, V., *Acta Paediatr. Scand. Suppl.*, 290, 1981. With permission.)

muscle cell cultures shown in earlier studies[62] seem to be associated with an increased serum HDL concentration.

VIII. CHANGE IN HABITUAL PHYSICAL ACTIVITY AND ITS RELATION TO SERUM LIPIDS

Acute physical exercise raised the serum triglycerides in 13 young elite athletes significantly[68] but did not cause significant changes in the TC or serum lipoprotein electrophoresis. Essentially similar results were seen in Kindermann's study,[69] whose lipid values were measured before and after a 10 km competitive run (11 to 14 year old boys).

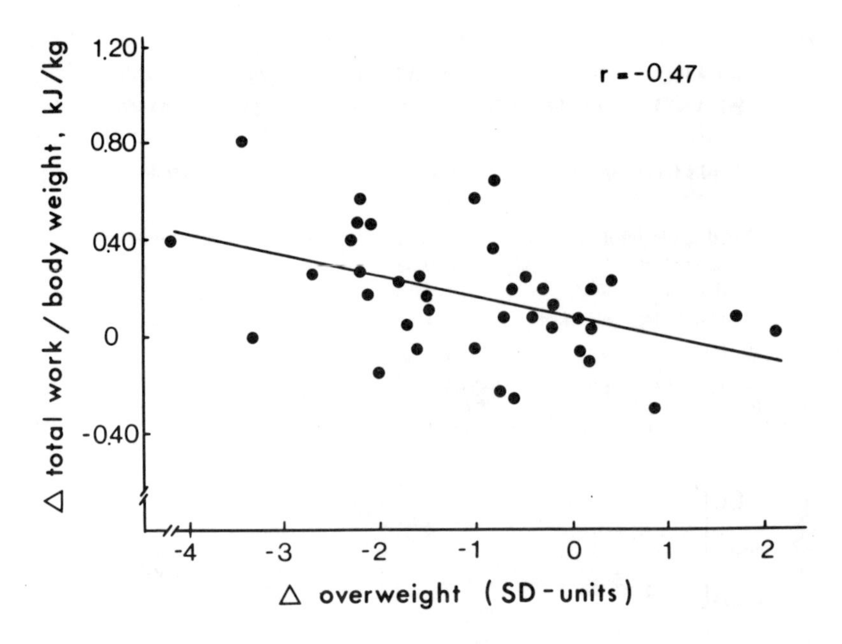

FIGURE 4. Correlation between change in total work/body weight and change in overweight (r = −0.47, $p<0.01$). Overweight is given in units of standard deviation.

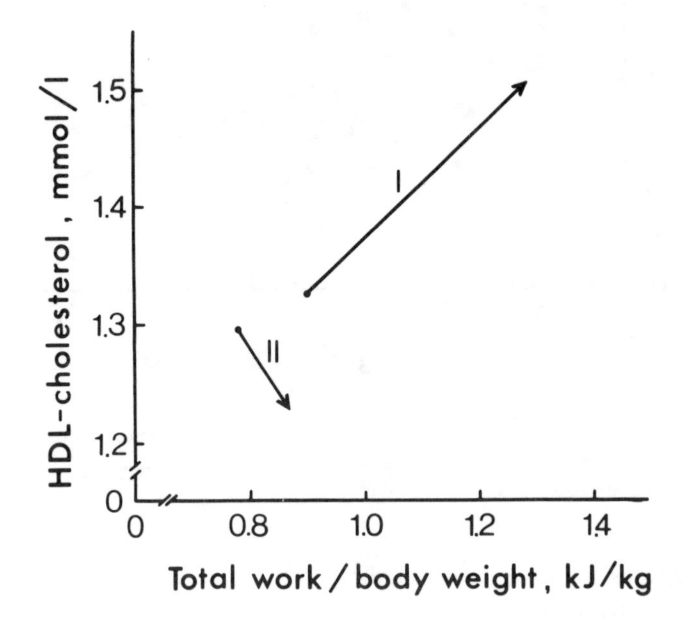

FIGURE 5. Correlation between HDL-cholesterol and mean change in total work/body weight. I = group with weight reduction, II = group with no weight reduction.

Gilliam and Burke[48] trained 14 girls (aged 8 to 10 years) for 6 weeks, five times per week, using a specially designed training program lasting 40 min. The TC decreased slightly, but not significantly, the HDL cholesterol increased significantly, and this caused a significant rise in the HDL/TC ratio. The knowledge of blood abnormalities can motivate changes in the lifestyle to make for better health and reduce the risk of

CAD. This was demonstrated in a study[70] where the changes in serum lipids and the reasons for these changes were evaluated in 108 medical students. The largest decreases in lipid values were found in those who changed their diet and exercise habits.

In a short duration study of the effect of diet and exercise in obese children (11 to 13 years) Widhalm et al.[71] showed that weight reduction was associated with a significant decrease of the LDL cholesterol, whereas the HDL cholesterol remained unchanged. Vigorous walking also induced significant changes in the lipoprotein levels in obese young men.[72] The subjects expended about 1100 kcal per session five times weekly during 16 weeks. The TC and triglycerides decreased but not significantly. The LDL cholesterol decreased and the HDL cholesterol increased, both significantly resulting in emphatic improvement of the HDL/LDL ratio (Table 2).

In another study of 61 obese school-children it was found that the physical performance correlated negatively with the degree of overweight (Figure 3) and that a reduction of the obesity after 2 years of treatment was followed by a statistically significant increase in the physical performance (Figure 4).[28]

In one group of children, where the children's weight was reduced to normal or slight overweight after 2 years of treatment, the physical performance and HDL cholesterol increased significantly whereas in the others with no weight reduction the physical performance increased slightly and the HDL cholesterol decreased (Figure 5).[28]

IX. CONCLUSIONS

It can be concluded that

1. Children on a high level of physical activity differ from sedentary children in several respects. The LDL cholesterol is lower. The HDL cholesterol is higher. The HDL/TC is higher. The triglyceride level is lower.
2. These differences cannot always be demonstrated and are more often seen in boys than in girls.
3. The differences are smaller in children and adolescents than in adults. This is probably because of the smaller variation in physical activity in childhood.
4. The experimental design and multifactorial interrelations between habitual activity, physical fitness, nutritional status, growth, and serum lipids must be taken into account when the lipid patterns are related to physical training in children.
5. Acute physical exercise has only slight effects on the serum cholesterol and triglyceride levels.
6. Rather short training programs can induce beneficial changes in the lipid values, especially in obese subjects.
7. It seems important to keep the level of physical activity of children high, because the basis of the level of habitual physical activity is laid during the childhood.
8. Long-term training programs in obese children reduce the degree of overweight and have positive effects on the lipid metabolism and physical performance capacity.

REFERENCES

1. **Strong, W. B.,** Is atherosclerosis a pediatric problem: an overview, in *Atherosclerosis, its Pediatric Aspects,* Strong, W. B., Ed., Grune & Stratton, New York, 1978, 1.
2. **Huttunen, J. K., Länsimies, E., Voutilainen, E., Ehnholm, C., Hietanen, E., Penttilä, I., Siitonen, O., and Rauramaa, R.,** Effect of moderate physical exercise on serum lipoproteins. A controlled clinical trial with special reference to serum high-density lipoproteins, *Circulation,* 60, 1220, 1979.
3. **Lehtonen, A. and Viikari, J.,** Serum triglycerides and cholesterol and serum high-density lipoprotein cholesterol in highly physical active men, *Acta Med. Scand.,* 204, 111, 1978.
4. **Lehtonen, A. and Viikari, J.,** The effect of vigorous physical activity at work on serum lipids with a special reference to serum high-density lipoprotein cholesterol, *Acta Physiol. Scand.,* 104, 117, 1978.
5. **Makurat, D. and Rzokiecka-Koska, M.,** Effects of physical activity sex and age on cholesterol distribution between beta (LDL) and alpha (HDL) lipoprotein fractions, *Polski Tygodnik Lekarski,* 34, 821, 1979.
6. **Punsar, S. and Karvonen, M.,** Physical activity and coronary heart disease in populations from East and West Finland, *Adv. Cardiol.,* 18, 196, 1976.
7. **Hardell, L. I.,** Serum Lipids and Lipoproteins at Birth and in Early Childhood, Academic dissertation, Uppsala University, Sweden, 1978.
8. **Impivaara, O. and Rimpelä, M.,** The levels of umbilical cord blood cholesterol and triglycerides in Finland, *Duodecim,* 89, 899, 1973.
9. **Srinivasan, S. R., Frerichs, R. R., and Berenson, G. S.,** Serum lipids and lipoproteins in children, in *Atherosclerosis, its Pediatric Aspects,* Strong, W. B., Ed., Grune & Stratton, New York, 1978, 85.
10. **Ginsburg, B.-E.,** Studies on Lipoprotein and Cholesterol Metabolism in Newborn Infants, Academic dissertation, Karolinska Institute, Stockholm, Sweden, 1980.
11. **Darmady, J. M., Fosbrooke, A. S., and Lloyd, J. K.,** Prospective study of serum cholesterol levels during first year of life, *Br. Med. J.,* 2, 685, 1972.
12. **Ziegler, E. E. and Foman, S. J.,** Infant feeding and blood lipid levels during childhood, in *Childhood Prevention of Atherosclerosis and Hypertension,* Lauer, R. M. and Shekelle, R. B., Eds., Raven Press, New York, 1980, 121.
13. **Berenson, G. S., Srinivasan, S. R., and Webber, L. S.,** Prognostic significance of lipid profiles in children, in *Childhood Prevention of Atherosclerosis and Hypertension,* Lauer, R. M. and Shekelle, R. B., Eds., Raven Press, New York, 1980, 75.
14. **Irjala, K., Viikari, J., Peltola, O., Klemola, K., and Erkkola, R.,** Serum Lipids in Urban Finnish Children, presented at 3rd Scand. Symp. Atherosclerosis Res., Abstr., Turku, Finland, 1979, 29.
15. **Knuiman, J. T., Hermus, J. J., and Hautvast, J. G. A. J.,** Serum total and high density lipoprotein (HDL) cholesterol concentration in rural and urban boys from 16 countries, *Atherosclerosis,* 36, 529, 1980.
16. **Stamler, J.,** Improved life styles: their potential for the primary prevention of atherosclerosis and hypertension in childhood, in *Childhood Prevention of Atherosclerosis and Hypertension,* Lauer, R. M. and Shekelle, R. B., Eds., Raven Press, New York, 1980, 3.
17. **Williams, P., Robinson, D., and Bailey, A.,** High-density lipoprotein and coronary risk factors in normal men, *Lancet,* 1, 72, 1979.
18. **Orchard, T. J., Rodgers, M., Hedley, A. J., and Mitchell, J. R. A.,** Changes in blood lipids and blood pressure during adolescence, *Br. Med. J.,* 2, 1563, 1980.
19. **van der Haar, F. and Kromhout, D.,** Food Intake, Nutritional Anthropometry and Blood Chemical Parameters in Three Selected Dutch Schoolchildren Populations, Academic dissertation, Agricultural University, Wageningen, The Netherlands, 1978.
20. **Fredrickson, D. S., Goldstein, J. L., and Brown, M. S.,** The familiar hyperlipoproteinemias, in *The Metabolic Basis of Inherited Disease,* 4th ed., Stanbury, J. B., Wyngaarden, J. B., and Fredrickson, D. S., Eds., McGraw-Hill, New York, 1978, 604.
21. **Nikkilä, E. A.,** Metabolic and endocrine control of plasma high density lipoprotein concentration, in *High Density Lipoproteins and Atherosclerosis,* Gotto, A. M., Jr., Miller, N. E., and Oliver, M. F., Eds., Elsevier/North-Holland Biomedical Press, Amsterdam, 1978, 177.
22. **Havel, R. J., Eder, H. A., and Bragdon, J. H.,** The distribution and chemical composition of ultracentrifugally separated lipoproteins in human serum, *J. Clin. Invest.,* 34, 1345, 1955.
23. **Viikari, J. and Saarni, H.,** Faster lipoprotein isolation by gradient flotation, *Clin. Chem.,* 24, 174, 1978.
24. **Cobb, S. A. and Sanders, J. L.,** Enzymic determination of cholesterol in serum lipoproteins separated by electrophoresis, *Clin. Chem.,* 24, 1116, 1978.

25. **Warnick, G. R., Cheung, M. C., and Albers, J. J.,** Comparison of current methods for high-density lipoprotein cholesterol quantitation, *Clin. Chem.,* 25, 596, 1979.
26. **Clarke, W. R., Schrott, H. G., Leaverton, P. E., Connor, W. E., and Lauer, R. M.,** Tracking of blood lipids and blood pressures in school age children: the Muscatine Study, *Circulation,* 58, 626, 1978.
27. **Laskarzewski, P., Morrison, J. A., deGroot, I., Kelly, K. A., Mellies, M. J., Khoury, P., and Glueck, C. J.,** Lipid and lipoprotein tracking in 108 children over a four-year period, *Pediatrics,* 64, 584, 1979.
28. **Ylitalo, V.,** Treatment of obese schoolchildren with special reference to the mode of therapy, cardiorespiratory performance and the carbohydrate and lipid metabolism, *Acta Paediatr. Scand. Suppl.,* 290, 1981.
29. **Harno, K.,** Metabolic Abnormalities in Siblings of Patients with Obesity and Maturity Onset Diabetes—A Study of Glucose Tolerance, Plasma Insulin, Insulin Binding and Serum Lipoproteins, Academic dissertation, Helsinki University, Finland, 1980.
30. **Montoye, H. J., Epstein, F. H., and Kjelsberg, M. O.,** Relationship between serum cholesterol and body fatness, *Am. J. Clin. Nutr.,* 18, 397, 1966.
31. **Bullen, B. A., Reed, R. B., and Mayer, J.,** Physical activity of obese and nonobese adolescent girls appraised by motion picture sampling, *Am. J. Clin. Nutr.,* 14, 211, 1964.
32. **Saris, W. H. M. and Binkhorst, R. A.,** The use of pedometer and actometer in studying daily physical activity in man. I. Reliability of pedometer and actometer, *Eur. J. Appl. Physiol.,* 37, 219, 1977.
33. **Holter, N. J.,** New method for heart studies, *Science,* 134, 1214, 1961.
34. **Telama, R.,** On methods for measuring physical activity, *Ser. Inst. Educ. Res. Univ. Jyväskylä Finland,* 46, 1, 1968.
35. **Elo, O., Hirvonen, L., Peltonen, T., and Välimäki, I.,** Physical working capacity of normal and diabetic children, *Ann. Paediatr. Fenn.,*11, 25, 1965.
36. **Eriksson, B. O. and Koch, G.,** Effect of physical training on hemodynamic response during submaximal and maximal exercise in 11–13 year old boys, *Acta Physiol.,* 87, 27, 1973.
37. **Thoren, C.,** Exercise testing in children, *Paediatrician,* 7, 100, 1978.
38. **Adams, F. H., Bengtsson, E., Berven, H., and Wegelius, C.,** The physical working capacity of normal schoolchildren. II. Swedish city and country, *Pediatrics,* 28, 243, 1961.
39. **Wahlund, H.,** Determination of the physical working capacity, *Acta Med. Scand. Suppl.,* 215, 1948.
40. **Mocellin, R., Lindemann, H., Rutenfranz, J., and Sbresny, W.,** Determination of W_{170} and maximal oxygen uptake in children by different methods, *Acta Paediatr. Scand. Suppl.,* 217, 13, 1971.
41. **Arstila, M.,** Pulse-conducted triangular exercise-ECG test. A feedback system regulating work during exercise, *Acta Med. Scand. Suppl.,* 529, 1972.
42. **Välimäki, I., Petäjoki, M.-L., Arstila, M., Viherä, P., and Wendelin, H.,** Automatically controlled ergometer for pulse-conducted exercise-test, *Med. Sport,* 11, 47, 1978.
43. **Petäjoki, M.-L., Arstila, M., and Välimäki, I.,** Pulse-conducted exercise test in children, *Acta Paediatr., Belg. Suppl.,* 28, 40, 1974.
44. **Hursti, M.-L., Pihlakoski, L., Antila, K., Halkola, L., and Välimäki, I.,** Experience of pulse-conducted ergometry in trained, normal and physically inactive schoolchildren, in *Children and Exercise,* Vol. 9, Int. Ser. Sport Sci. 10, Berg, K. and Eriksson, B. O., Eds., University Park Press, Baltimore, 1980, 128.
45. **Välimäki, I., Hursti, M.-L., Pihlakoski, L., Wanne, O., Halkola, L., and Viikari, J.,** Relation of serum lipids to ergometric performance in school children, in *Physical Training in Health Promotion and Medical Care,* Abstr., Publications of the University of Kuopio, 1, 83, 1980.
46. **Välimäki, I., Hursti, M.-L., Pihlakoski, L., and Viikari, J.,** Exercise performance and serum lipids in relation to physical activity in schoolchildren, *Int. J. Sports Med.,* 1, 132, 1980.
47. **Ekblom, B.,** Effect of physical training in adolescent boys, *J. Appl. Physiol.,* 27, 350, 1969.
48. **Gilliam, T. B. and Burke, M. B.,** Effects of exercise on serum lipids and lipoproteins in girls, ages 8 to 10 years, *Artery,* 4, 203, 1978.
49. **Lussier, L. and Buskirk, E. R.,** Effects of an endurance training regimen on assesment of work capacity in prepubertal children, *Ann. N.Y. Acad. Sci.,* 301, 734, 1977.
50. **Borms, J., Hebbelinck, M., and Duquet, W.,** On the variability of some physical fitness parameters in boys 6–13 years of age, in *Physical Fitness,* Seliger, V., Ed., Universita Karlova, Praha, 1973, 91.
51. **Johnson, M. L., Burke, B. S., and Mayer, J.,** Relative importance of inactivity and overeating in the energy balance of obese high school girls, *Am. J. Clin. Nutr.,* 4, 37, 1956.
52. **Stefanik, P. A., Heald, F. P., Jr., and Mayer, J.,** Caloric intake in relation to energy output of nonobese and obese adolescent boys, *Am. J. Clin. Nutr.,* 7, 55, 1959.
53. **Rose, H. E. and Mayer, J.,** Activity, calorie intake, fat storage and the energy balance of infants, *Pediatrics,* 41, 18, 1968.

54. **Mayer, J.,** Nutrition, in *Proc. VIII Int. Congr.*, Masek, J., Osancova, K., and Cuthbertson, D. B., Eds., Excerpta Medica, Amsterdam 1970, 354.
55. **Shephard, R. J.,** The working capacity of schoolchildren, in *Frontiers of Fitness*, Shephard, R. J., Ed., Charles C. Thomas, Springfield, Ill., 1971, 319.
56. **Pařizková, J.,** Consequences of adaptation to increased physical activity in obese children, in *Body Fat and Physical Fitness*, Pařizková, J., Ed., Martinus Nijhoff, 1977, 169.
57. **Börjeson, M.,** Overweight children, *Acta Paediatr. Scand. Suppl.*, 51, 132, 1962.
58. **Rehs, H.-J., Berndt, I., and Rutenfranz, J.,** Untersuchungen zur Frage der Leistungsfähigkeit Adipöser unter besonderer Berücksictigung des Sportunterrichtes, *Z. Kinderheilkunde*, 115, 23, 1973.
59. **Schwarz, W. and Huber, E. G.,** Erste Ergebnisse kindersportärzlicher Untersuchungen Normwerte und Einfluss des übergewichts auf die körperliche Leistungsfähigkeit, *Päd. Pädol*, 12, 106, 1977.
60. **Wood, P. D., Haskett, W., Klein, H., Lewis, S., Stern, M. P., and Farquhar, J.,** The distribution of plasma lipoproteins in middle-aged runners, *Metabolism*, 25, 1249, 1976.
61. **Ismail, A. H. and Montgomery, D. L.,** The effect of a four-month physical fitness program on a young and an old group matched for physical fitness, *Eur. J. Appl. Physiol.*, 40, 137, 1979.
62. **Tammi, M., Rönnemaa, T., Vihersaari, T., Saarni, H., Lehtonen, A., and Viikari, J.,** Effect of hyperlipidemia and HDL-emic human sera on the synthesis of DNA and glycosaminoglycans by cultured human aortic smooth muscle cells, *Med. Biol.*, 57, 118, 1979.
63. **Lee, C. J.,** Nutritional status of selected teenagers in Kentucky, *Am. J. Clin. Nutr.*, 31, 1453, 1978.
64. **Schwane, J. A. and Cundiff, D. E.,** Relationship among cardiorespiratory fitness, regular physical activity, and plasma lipids in young adults, *Metabolism*, 28, 771, 1979.
65. **Gilliam, T. B., Katch, V. L., Thorland, W., and Weltman, A.,** Prevalence of coronary heart disease risk factors in active children, 7 to 12 years of age, *Med. Sci. Sports*, 9, 21, 1977.
66. **Montoye, H. J., Block, W. D., and Gayle, R.,** Maximal oxygen uptake and blood lipids, *J. Chron. Dis.*, 31, 111, 1978.
67. **Rönnemaa, T., Lehtonen, A., Järveläinen, H., and Viikari, J.,** Plasma lipids and lipoproteins of young male athletes and the effect of their sera on cultured human aortic smooth muscle cells, *Scand. J. Sports. Sci.*, 2, 33, 1980.
68. **Novosad, P., Hofhanzl, C., Vaskova, I., Rybka, J., and Lanicek, P.,** Serum lipid changes following maximum anaerobic exercise in relation to some other biochemical parameters, *Cas. Lek. Ces.*, 118, 744, 1979.
69. **Kindermann, W., Keul, J., and Lehmann, M.,** Ausdauerbelastungen beim Heranwachsenden-metabolische und kardiozirkulatorische Veränderungen. Stoffwechsel- und Herz-Kreislauf-Veränderungen bei einem 10-km-Wettkampflauf 11- bis 14-jähriger Jungen, *Fortschr. Med.*, 97, 659, 1979.
70. **Yates, B., Johnson, W. D., Wingo, C., and Lopez, S. A.,** Serum lipid changes in medical students, *J. Am. Dietet. Assoc.*, 72, 398, 1978.
71. **Widhalm, K., Maxa, L., and Zyman, H.,** Effect of diet and exercise upon the cholesterol and triglyceride content of plasma lipoproteins in overweight children, *Eur. J. Pediatr.*, 127, 121, 1978.
72. **Leon, A. S., Conrad, J., Hunninghake, D. B., and Serfass, R.,** Effects of a vigorous walking program on body composition and carbohydrate and lipid metabolism of obese young men, *Am. J. Clin. Nutr.*, 33, 1776, 1979.

Metabolic and Cardiovascular Diseases and Exercise

Chapter 14

PHYSICAL TRAINING IN OBESITY AND DIABETES MELLITUS

Katriina Kukkonen, Rainer Rauramaa, and Esko Länsimies

TABLE OF CONTENTS

I. INTRODUCTION

Disorders of lipid metabolism, which occur in various forms in clinical disease entities such as obesity and diabetes, bear a substantial risk for development of coronary heart disease. Obesity in the middle-aged is fairly common and often associated with impaired glucose tolerance or manifest noninsulin dependent diabetes mellitus as well as with low level of high density lipoproteins in serum. Such an accumulation of several risk factors for coronary heart disease may even be accentuated by the presence of increased blood pressure, which on the other hand, has even been suggested to be mediated through disturbed carbohydrate metabolism.[1]

Obesity-associated risk factors are often, however, reversible by reducing the amount of adipose tissue. This may be achieved with the aid of dietary manipulation but it is not always possible to normalize decreased high density lipoprotein (HDL) even if very low density lipoprotein (VLDL) can be decreased.[2] Physical training is a physiological approach and is well documented in healthy persons to have a favorable effect as regards body composition[3] as well as serum lipoproteins.[4]

II. OBESITY

A. General

Obesity means increased amount of adipose tissue. At both extreme ends of body weight (anorexia nervosa and morbid obesity) it is fairly easy to recognize energy imbalance, but in the majority of cases the diagnosis of obesity is based on statistical or operational (i.e., morbidity and mortality) definitions.[5] The implication is that the prevalence of obesity in one population is not necessarily directly comparable to another. Therefore, the criteria of obesity applied also affect the indications for therapy. Physical training is one option among the treatment modalities for obesity and most often used in nonmorbid obesity.

At the cellular level, obesity has been attempted to be characterized according to size and amount of adipocytes. The existence of two types of obesity, based on fat cell number and size, has been proposed[6–8] and it has also been used for prognosis in obesity treatment.[9] However, the concept of hyperplastic obesity has been questioned by Jung et al.[10] and Garrow[11] as well as its prognostic significance.[12]

Increased amount of adipocytes is also associated with various endocrine derangements.[13,14] Obese persons have often hyperinsulinemia[15] both at rest and during exercise.[13] In alignment to animal models Kopelman et al.[14] found two different types of endocrine responses to symptomatic hypoglycemia (intravenous insulin tolerance test) in humans suggesting for hypothalamic (genetic, obesity starting early in childhood) and acquired obesity. Obesity has also been suggested to be due to impaired regulation of thermogenesis.[16–19]

At this moment, when evaluating different aspects of obesity treatment studies, it is worthwhile to remember the possible different types of obesity and the possible differences in the morphological and biochemical properties in obesity which influence exercise-induced changes in lipid metabolism.

Increased plasma levels of lipoproteins may be due to increased production or decreased clearance.[15] Despite the mechanisms, total plasma triglycerides and cholesterol are often increased in obesity;[20] low density lipoprotein (LDL) and very low density lipoprotein (VLDL) are increased while high density lipoprotein (HDL) is decreased.

In many population studies, HDL has been shown to be inversely correlated with body weight.[21–24] LDL cholesterol closely parallels total cholesterol, and VLDL triglycerides bear an inverse relationship to HDL cholesterol. One possible mediator in lipoprotein metabolism is supposed to be composed of different triglyceride lipases (vascular endothelial and hepatic), which hydrolyze plasma triglycerides. This facilitates the transport of fatty acids to adipose and muscle tissues where they are again stored or used for energy.

In addition to the key role of insulin[25] in the regulation of lipoprotein lipase, nutrition,[26] exercise,[27,28] and training state[29] are important contributors. The lipolytic activity of adipose tissue varies from site to site in human body being highest in gluteal and femoral adipose tissue.[30] Acute fasting decreases lipoprotein lipase activity in adipose tissue[31] but after obese people have lost weight and body weight has stabilized, this lipolytic activity has been shown to increase paradoxically.[32]

B. Effect of Exercise on Body Composition and Serum Lipids

In normal weight persons increased physical activity usually decreases adipose tissue and body weight and increases fat-free tissue.[33] This has also been found to take place in obese persons after physical training, the relative changes being even more significant.[34–36] However, hyperplastic obese persons do not seem to reduce weight but on the contrary adipose tissue may even increase after training.[1,37–39]

After weight reduction, the size of fat cells decreases but the number of the cells does not change substantially. Based on a fairly large group of subjects Krotkiewski et al.[9] have suggested that fat cell number is a major determinant in the success of weight reduction, at least after dietary treatment. This means that hyperplastic obese persons fare worse when compared to the hypertrophic obese. However, Ashwell et al.[12] emphasize the prognostic importance of basal metabolic rate.

There are few studies concerning the effect of physical training on serum lipoproteins in the obese. Since all of these studies have been uncontrolled with rather few subjects, conclusions on plasma lipids and physical training in obesity must be very cautious. In addition, one should take into consideration other possible changes in lifestyle (diet, alcohol, smoking) concomitant with increased leisure time physical activity.

Lewis et al.[40] trained 22 obese middle-aged women for 17 weeks. The training program consisted of supervised jogging twice a week for 20 min at the intensity of 80% of maximal heart rate as well as calisthenics twice a week for 1 hr. There was a significant decrease in body fat, but no changes in cholesterol or in its lipoprotein fractions. However, HDL-LDL cholesterol ratio increased.

Leon et al.[35] recruited six obese young men who were trained under supervision in laboratory for 16 weeks. The exercise program consisted of vigorous walking (3.2 mi/hr) on treadmill for 90 min five times per week. Training had no effect on plasma cholesterol and triglycerides while both HDL cholesterol and HDL-LDL cholesterol ratio increased significantly. These improvements in cholesterol and its lipoprotein fractions are in agreement with results obtained from nonobese middle-aged men after moderate training.[4] The studies of Lewis et al.[40] and Leon et al.[35] are encouraging in indicating the potential effects of physical training in modifying serum lipoproteins with respect to atherosclerosis.

Although muscle work has no effect on either normal or abnormal fasting blood glucose in the obese, it results in increased peripheral insulin sensitivity. This has been clearly demonstrated both during physical exercise and after physical training,[37,38,41] suggestive of the beneficial contribution of training to health.

Obesity is often combined with dyslipoproteinemias of which the types II and IV are the most common. Although physical training is accepted as part of the therapy to reduce overweight, there are again few studies available. These are short-term uncontrolled studies with few subjects. In type II dyslipoproteinemia, LDL cholesterol decreased after 10 weeks'[42] or 6 months'[43] training. HDL cholesterol remained unchanged[43] or increased.[42] Originally, normal triglycerides decreased somewhat in both studies.

Lampman et al.[44–46] have applied physical training with varying intensity combined with diet in type IV dyslipoproteinemia. Without exceptions, these regimens have resulted in decreased triglycerides which could not be explained with a concomitant decrease of overweight. Exercise-induced decrease in triglycerides occurred somewhat later than when using only diet but was maintained longer. Similar results of decrease in VLDL triglycerides have also been reported by Oscai et al.,[47] Gyntelberg et al.,[48] and by Giese et al.[49] after only 4 days of exercise. This implies the potential and rapidly occurring effect of physical activity, which also vanishes in a few days. The beneficial effects of physical training are also advocated by the fact that even if diet normalizes VLD lipoproteins it may not always have effect on low HD lipoproteins in type IV dyslipoproteinemia.[2] Since the ratio of HDL cholesterol and total cholesterol is probably the best parameter to describe the risk of coronary heart disease, the promotion of increased physical activity is also beneficial in various types of dyslipoproteinemias.

III. DIABETES MELLITUS

A. General

Diabetes mellitus is characterized by chronic elevation of blood glucose concentrations. Accordingly, the diagnosis of diabetes also rests on the determination of blood glucose level, either fasting and/or 2 hr after standard glucose load. For diagnostic purposes, new reference values have been established for diabetes mellitus and also for the so-called impaired glucose tolerance.[50,51]

Glucose homeostasis is achieved by delicate endocrine control of insulin and its counteracting hormones. Insulin is by far the most important hormonal determinant of blood glucose concentration. Human diabetes is characterized by disturbed plasma insulin level. Insulin dependent (IDDM) or type 1 diabetes is due to diminished endogenous insulin secretion while noninsulin dependent type (NIDDM) or type 2 diabetes is principally described as a state of hyperinsulinism. Type 2 diabetics may be either nonobese or obese, as are also persons with impaired glucose tolerance.

Except for regulating blood glucose homeostasis, insulin plays a major role in the control of lipid metabolism. From many epidemiologic data it is obvious that diabetic subjects are exceptionally prone to macro- or microangiopathic vascular complications. Macroangiopathic lesions are found in coronary, cerebral, and lower limb arteries and are not dependent on the duration or severity of diabetes or on blood glucose concentrations. On the other hand, microangiopathic complications (retinopathy and nephropathy) are closely linked to blood glucose balance over long time periods.

In population studies, hyperglycemia has been associated with the development of atherosclerosis. In the Whitehall study[52] impaired glucose tolerance nearly doubled coronary heart disease mortality compared with subjects without disturbances in glucose tolerance. Besides increased risk for coronary heart disease in the presence of other risk factors, impaired glucose tolerance may progress to manifest diabetes, even if slowly.[53] However, Stamler et al.[54] were unable to find any association between glucose intolerance and coronary heart disease.

Data from the Framingham study suggest that diabetes seems to have some unique effects on the development of coronary heart disease, especially in women, which cannot be explained by the usual risk factors.[55] Decreasing blood glucose to the so-called normal level does not affect the later development of vascular complications. Therefore, it has been suggested that disturbances in plasma insulin level may be associated with vascular complications in a way that is so far undefined. Insulin treated diabetic subjects are not actually insulin deficient but on the contrary have supraphysiological amounts of exogenous insulin in circulation unrelated to instantaneous nutritional requirements.[56] Noninsulin dependent diabetic subjects, most of whom are obese, have often increased basal level of insulin.

In population studies, high level of plasma insulin has been shown to be an independent risk factor associated with coronary heart disease.[57,58] Physical exercise[13] or physical training[37,38,41] in the obese has been shown to decrease plasma insulin in resting state or during glucose challenge. The present data on insulin and lipid disturbances in diabetes may justify the use of physical training as one method of improving metabolic balance.

B. Serum Cholesterol in Diabetes Mellitus

Generally diabetic patients show increased levels of serum cholesterol and its LDL fraction. Data on HDL cholesterol are somewhat controversial and in this respect the type of diabetes has to be taken into consideration. This topic has recently been reviewed by Witztum and Schonfeld.[59]

In insulin dependent diabetic (IDDM) subjects total cholesterol[60–63] or LDL cholesterol[63,64] are similar to nondiabetic persons. However, VLDL cholesterol, which is a minor fraction of cholesterol, has been reported to be slightly higher in IDDM.[63] HDL cholesterol is either equal[61,63,65] or greater[60,66,67] in IDDM compared to nondiabetic subjects. Increased HDL level in diabetics seems to contradict with epidemiologic data on the high prevalence of atherosclerosis in diabetes mellitus. This disparity has been explained by alterations in HDL composition (HDL_2 and HDL_3 subfractions, and apolipoprotein A). Indeed, it has been suggested that elevation of HDL cholesterol in IDDM is due to increased HDL_3 subfraction,[68] which is not regarded antiatherogenic. Taskinen and Nikkilä[69] did not find abnormal lipoprotein structure. On the other hand, Reckless et al.[70] suggest that HD-lipoproteins are of less importance than LD-lipoproteins in relation to large vessel disease in diabetic subjects.

Actually, to achieve good control of IDDM requires supraphysiologic amounts of exogenous circulating insulin which has been suggested to be a central factor in atherogenesis.[56] Poor control of diabetes is associated with abnormally high levels of total cholesterol as well as of LDL and VLDL cholesterol.[63] Insulin treatment for 2 weeks decreased LD- and VLD-lipoproteins but had no effect on low HDL cholesterol.[69] In some studies HDL cholesterol has been shown to correlate negatively with glycosylated hemoglobin,[71] which reflects glucose balance in blood over several weeks, while Elkeles et al.[72] and Kennedy et al.[61] found no correlation.

In noninsulin dependent diabetes (NIDDM), when evaluating plasma lipid levels, oral antidiabetic medication and obesity may be confounding factors.[73] Total cholesterol is not different from nondiabetic subjects.[60,61,71,74] Respective similarity has also been shown for LDL and VLDL cholesterol.[66,74] On the other hand, a disparity exists with HDL cholesterol since it has been shown to be either similar[60,66,74] or decreased,[61,71] when compared to nondiabetic subjects. Lopes-Virella et al.[71] found no abnormalities in apolipoprotein A.

C. Serum Triglycerides in Diabetes Mellitus

Hypertriglyceridemia is a common dyslipoproteinemia. Carlson and Böttiger[75] have proposed hypertriglyceridemia to be an independent risk factor for coronary heart disease, but newer analyses of epidemiologic studies[76] do not reveal a causality.

In untreated[69] and ketotic[77] insulin dependent diabetics, as well as in patients in poor control of the disease,[63] elevated levels of triglycerides have been found in all lipoprotein fractions except HDL,[69] which carries only a minor proportion of triglycerides.

Good control of diabetes evaluated with the aid of glycosylated hemoglobin was accompanied with serum triglycerides concentration not different from nondiabetic siblings.[63] In noninsulin dependent diabetes mellitus, serum triglycerides are either similar[74] or increased[64,66,71] when compared to healthy persons.

Several mechanisms have been postulated for the elevation of triglycerides, which seems to be due to a combined effect of increased hepatic secretion of VLDL triglycerides and their decreased removal from plasma.[77] The former process is insulin dependent while in the latter the activity of lipoprotein lipase is essential, though at present detailed data are still deficient in human diabetes.

D. Physical Exercise in Diabetes Mellitus

The therapeutic use of physical exercise in diabetes has a tradition of several centuries, extending far before the use of diet, drugs, and exogenous insulin. At present the influence of exercise on glucose homeostasis and substrate utilization in energy metabolism are known rather extensively.[78,79] Of crucial importance during physical exercise is the maintenance of glucose homeostasis. In healthy subjects this is achieved with adjustments of insulin secretion.[80] Noninsulin dependent patients mostly respond to exercise not differently from healthy persons.

In insulin dependent diabetic patients in at least moderate glucose balance, exercise alleviates hyperglycemia while in ketotic patients exercise has a tendency to aggravate the disease.[81] The first mentioned change is a physiological response but the last mentioned disorder is due to absolute insulin deficiency. In some cases exercise may provoke hypoglycemia in diabetics, because unlike in healthy persons, plasma insulin level does not decrease during exercise in diabetic patients treated with exogenous insulin. This leads to a suppression of hepatic output of glucose and an increase in its peripheral utilization. Therefore, in spite of many beneficial effects of increased physical activity, the possible metabolic and cardiovascular risks of physical exercise in diabetic subjects always have to be taken into consideration.

It has been suggested that the site of exogenous insulin injection is important in connection with its absorption during exercise.[82] which has also been acknowledged in practice.[83] On the other hand, Kemmer et al.[84] did not find any increased absorption from the sites of injection involved in movement during exercise but instead they observed increased plasma insulin level after exercise. Discrepancies regarding insulin absorption during exercise remain to be clarified, but may be due, at least partly, to differences in experimental protocols used. Anyhow, when evaluating the metabolic responses of exercise and physical training, the effect of circulating insulin has to be evaluated carefully.

Plasma insulin is also the key regulator of lipolysis. In healthy subjects during exercise, insulin decreases with ensuing increased lipolysis. The same holds true for diabetic patients in good control of the disease,[80,85] while due to larger free fatty acids (FFA) availability the leg uptake of FFA is greater in ketotic patients. Therefore, in patients with less good control of the disease the result is accelerated lipolysis and ketogenesis.

Hagan et al.[86] observed no changes in plasma triglycerides and FFA after light and moderate exercise in young boys with IDDM. Physical training on the bicycle ergometer for 6 months in subjects with NIDDM decreased significantly both triglycerides and cholesterol, although the study was uncontrolled and included only six subjects.[87] In healthy subjects there have been accumulating data from epidemiological studies[88–91] suggesting beneficial effects of increased physical activity on coronary heart disease. This may partly be mediated through increased levels of HDL cholesterol, which can be manipulated by exercise.[4] As to diabetic subjects, corresponding data are lacking.

Obviously, as earlier recognized,[87] further studies are needed on physical training and plasma lipids in diabetic subjects. More studies are required to define the intensity and type of training to induce and maintain possible favorable changes in plasma lipids. As long as these data are not available, physical training for both insulin dependent and nondependent diabetic patients has to be prescribed individually, remembering the possible metabolic and cardiovascular hazards of exercise.

IV. CONCLUSIONS

Lipid disturbances associated with obesity and noninsulin dependent diabetes are additional risks factors for coronary heart disease, while dyslipoproteinemia with other known risk factors in insulin dependent diabetes mellitus does not alone account for the proneness to atherosclerosis.[55] Moderate physical training has favorable effects on lipid and carbohydrate metabolism although concomitant weight reduction is often only moderate. However, even if risk factors can be modified with increased physical activity, there is no evidence of the effect on atherogenesis. On the other hand, when physical training prescription and follow-up are adequately performed, resulting improved physical fitness has concomitant positive influence on psychosocial quality of life which cannot be neglected.

REFERENCES

1. **Krotkiewski, M., Mandroukas, K., Sjöström, L., Sullivan, L., Wetterqvist, H., and Björntorp, P.,** Effects of long-term physical training on body fat, metabolism, and blood pressure in obesity, *Metabolism,* 28, 650, 1979.
2. **Witztum, J. L., Dillingham, M. A., Giese, W., Bateman, J., Diekman, C., Kammeyer Blaufuss, E., Weidman, S., and Schonfeld, G.,** Normalization of triglycerides in type IV hyperlipoproteinemia fails to correct low levels of high-density-lipoprotein cholesterol, *N. Engl. J. Med.,* 303, 907, 1980.
3. **Björntorp, P.,** Effects of exercise and physical training on carbohydrate and lipid metabolism in man, *Adv. Cardiol.,* 18, 158, 1976.
4. **Huttunen, J. K., Länsimies, E., Voutilainen, E., Ehnholm, C., Hietanen, E., Penttilä, I., Siitonen, O., and Rauramaa, R.,** Effect of moderate physical exercise on serum lipoproteins. A controlled clinical trial with special reference to serum high-density lipoproteins, *Circulation,* 60, 1220, 1979.
5. **Berger, M., Berchtold, P., Gries, A., and Zimmerman, H.,** Indications for the treatment of obesity, in *Recent Advances in Obesity Research: III,* Björntorp, P., Cairella, M., and Howard, A. N., Eds., J. Libbey & Co. Ltd., London, 1981, chap. 1.
6. **Björntorp, P. and Sjöström, L.,** Number and size of adipose tissue fat cells in relation to metabolism in human obesity, *Metabolism,* 20, 703, 1971.
7. **Sjöström, L.,** Fat cells and body weight, in *Obesity,* Stunkard, A. S., Ed., W. B. Saunders, Philadelphia, 1980, chap. 3.
8. **Sjöström, L. and Björntorp, P.,** Body composition and adipose tissue cellularity in human obesity, *Acta Med. Scand.,* 195, 201, 1974.
9. **Krotkiewski, M., Sjöström, L., Björntorp, P., Carlgren, G., Garellick, G., and Smith, U.,** Adipose tissue cellularity in relation to prognosis for weight reduction, *Int. J. Obes.,* 1, 395, 1977.
10. **Jung, R. T., Gurr, M. I., Robinson, M. P., and James, W. P. T.,** Does adipocyte hypercellularity in obesity exist?, *Br. Med. J.,* 2, 319, 1978.
11. **Garrow, J. S.,** *Energy Balance and Obesity in Man,* Elsevier/North-Holland Biomedical Press, Amsterdam, 1978, 132.
12. **Ashwell, M., Durrant, M., and Garrow, J. S.,** Does adipocyte cellularity or the age of onset of obesity influence the response to short-term inpatient treatment of obese women?, *Int. J. Obes.,* 2, 449, 1978.

13. **Bray, G. A., Whipp, B. J., Koyal, S. N., and Wasserman, K.,** Some respiratory and metabolic effects of exercise in moderately obese men, *Metabolism,* 26, 403, 1977.
14. **Kopelman, P. G., Pilkington, T. R. E., White, N., and Jeffcoate, S. L.,** Evidence of existence of two types of massive obesity, *Br. Med. J.,* 280, 82, 1980.
15. **Nestel, P. and Goldrick, B.,** Obesity: changes in lipid metabolism and the role of insulin, *Clin. Endocrinol. Metab.,* 5, 313, 1976.
16. Editorial, Do the lucky ones burn off their dietary excesses?, *Lancet,* 2, 115, 1979.
17. Editorial, Metabolic obesity?, *Br. Med. J.,* 282, 172, 1981.
18. **Himms-Hagen, J.,** Obesity may be due to a malfunctioning of brown fat, *Can. Med. Assoc. J.,* 121, 1361, 1979.
19. **De Luise, M., Blackburn, G. L., and Flier, J. S.,** Reduced activity of the red-cell sodium-potassium pump in human obesity, *N. Engl. J. Med.,* 303, 1017, 1980.
20. **Albrink, M. J., Krauss, R. M., Lindgren, F. T., von der Groeben, J., Pan, S., and Wood, P. D.,** Intercorrelations among plasma high density lipoprotein, obesity, and triglycerides in a normal population, *Lipids,* 15, 668, 1980.
21. **Avogaro, P., Cazzolato, G., Bon, G. B., Quinci, G. B., and Chinello, M.,** HDL-cholesterol, apolipoproteins A and B. Age and index body weight, *Atherosclerosis,* 31, 85, 1978.
22. **Glueck, C. J., Taylor, H. J., Jacobs, D., Morrison, J. A., Beaglehole, R., and Williams, O. D.,** Plasma high-density lipoprotein cholesterol: association with measurements of body mass. The Lipid Research Clinics Program Prevalence Study, *Circulation,* 62(Suppl. 4), IV-62, 1980.
23. **Gordon, T., Castelli, W. P., Hjortland, M., Kannel, W. B., and Dawber, T. R.,** High density lipoprotein as a protective factor against coronary heart disease, *Am. J. Med.,* 62, 707, 1977.
24. **Rhoads, G. G., Gulbrandsen, C. L., and Kagan, A.,** Serum lipoproteins and coronary heart disease in a population study of Hawaii Japanese men, *N. Engl. J. Med.,* 294, 293, 1976.
25. **Pykälistö, O. J., Smith, P. H., and Brunzell, J. D.,** Determinants of human adipose tissue lipoprotein lipase. Effect of diabetes and obesity on basal- and diet-induced activity, *J. Clin. Invest.,* 56, 1108, 1975.
26. **Tan, M. J.,** The lipoprotein lipase system: new understandings, *Can. Med. Assoc. J.,* 118, 675, 1978.
27. **Borensztajn, J., Rone, M. S., Babirak, S. P., McGarr, J. A., and Oscai, L. B.,** Effect of exercise on lipoprotein lipase activity in rat heart and skeletal muscle, *Am. J. Physiol.,* 229, 394, 1975.
28. **Lithell, H., Hellsing, K., Lundqvist, G., and Malmberg, P.,** Lipoprotein-lipase activity of human skeletal muscle and adipose tissue after intensive physical exercise, *Acta Physiol. Scand.,* 105, 312, 1979.
29. **Nikkilä, E. A., Taskinen, M.-R, Rehunen, S., and Härkönen, M.,** Lipoprotein lipase activity in adipose tissue and skeletal muscle of runners: relation to serum lipoproteins, *Metabolism,* 27, 1661, 1978.
30. **Lithell, H. and Boberg, J.,** The lipoprotein-lipase activity of adipose tissue from different sites in obese women and relationship to cell size, *Int. J. Obes.,* 2, 47, 1978.
31. **Huttunen, J. K., Ehnholm, C., Nikkilä, E. A., and Ohta, M.,** Effect of fasting on two postheparin plasma lipases and triglyceride removal in obese subjects, *Eur. J. Clin. Invest.,* 5, 435, 1975.
32. **Schwartz, R. S. and Brunzell, J. D.,** Increased adipose-tissue lipoprotein-lipase activity in moderately obese men after weight reduction, *Lancet,* 1, 1230, 1978.
33. **Pollock, M. L., Cureton, T. K., and Greninger, L.,** Effects of frequency of training on working capacity, cardiovascular function, and body composition of adult men, *Med. Sci. Sports,* 1, 70, 1969.
34. **Boileau, R. A., Buskirk, E. R., Horstman, D. H., Mendez, J., and Nicholas, W. C.,** Body composition changes in obese and lean men during physical conditioning, *Med. Sci. Sports,* 3, 183, 1971.
35. **Leon, A. S., Conrad, J., Hunninghake, D. B., and Serfass, R.,** Effects of a vigorous walking program on body composition, and carbohydrate and lipid metabolism of obese young men, *Am. J. Clin. Nutr.,* 32, 1776, 1979.
36. **Moody, D. L., Kollias, J., and Buskirk, E. R.,** The effect of a moderate exercise program on body weight and skinfold thickness in overweight college women, *Med. Sci. Sports,* 1, 75, 1969.
37. **Björntorp, P., de Jounge, K., Krotkiewski, M., Sullivan, L., Sjöström, L., and Stenberg, J.,** Physical training in human obesity. III. Effects of long-term physical training on body composition, *Metabolism,* 22, 1467, 1973.
38. **Björntorp, P., de Jounge, K., Sjöström, L., and Sullivan, L.,** The effect of physical training on insulin production in obesity, *Metabolism,* 19, 631, 1970.
39. **Sullivan, L.,** Metabolic and physiologic effects of physical training in hyperplastic obesity, *Scand. J. Rehabil. Med.,* Suppl. 5, 1, 1976.
40. **Lewis, S., Haskell, W. L., Wood, P. D., Manoogian, N., Bailey, J. E., and Pereira, M.,** Effects of physical activity on weight reduction in obese middle-aged women, *Am. J. Clin. Nutr.,* 29, 151, 1976.

41. **Björntorp, P., Holm, G., Jacobson, B., Schiller-de Jounge, K., Lundberg, P.-A., Sjöström, L., Smith, U., and Sullivan, L.,** Physical training in human hyperplastic obesity. IV. Effects on the hormonal status, *Metabolism,* 26, 319, 1977.
42. **Roundy, E. S., Fisher, G. A., and Anderson, A.,** Effect of exercise on serum lipids and lipoproteins, *Med. Sci. Sports,* 10 (Abstr.) 55, 1978.
43. **Melish, J., Bronstein, D., Gross, R., Dann, D., White, J., Hunt, H., and Brown, W. V.,** Effect of exercise in type II hyperlipoproteinemia, *Circulation,* 57(Suppl. 2, Abstr.), II-38, 1978.
44. **Lampman, R. M., Santinga, J. T., Bassett, D. R., Block, W. D., Mercer, N., Hook, D. A., Flora, J. D., and Foss, M. L.,** Type IV hyperlipoproteinemia: effects of a caloric restricted type IV diet versus physical training plus isocaloric type IV diet, *Am. J. Clin. Nutr.,* 33, 1233, 1980.
45. **Lampman, R. M., Santinga, J. T., Bassett, D. R., (MonDragon) Mercer, N., Block, W. D., Flora, J. D., Foss, M. L., and Thorland, W. G.,** Effectiveness of unsupervised and supervised high intensity physical training in normalizing serum lipids in men with type IV hyperlipoproteinemia, *Circulation,* 57, 172, 1978.
46. **Lampman, R. M., Santinga, J. T., (LaValley) Hodge, M. F., Block, W. D., Flora, J. D., and Bassett, D. R.,** Comparative effects of physical training and diet in normalizing serum lipids in men with type IV hyperlipoproteinemia, *Circulation,* 55, 652, 1977.
47. **Oscai, L. B., Patterson, J. A., Bogard, D. L., Beck, R. J., and Rothermel, B. L.,** Normalisation of serum triglycerides and lipoprotein electrophoretic patterns by exercise, *Am. J. Cardiol.,* 30, 775, 1972.
48. **Gyntelberg, F., Brennan, R., Holloszy, J. O., Schonfeld, G., Rennie, M. J., and Weidman, S. W.,** Plasma triglyceride lowering by exercise despite increased food intake in patients with type IV hyperlipoproteinemia, *Am. J. Clin. Nutr.,* 30, 716, 1977.
49. **Giese, M. D., Nagle, F. J., Corliss, R. J., and Westgard, J. O.,** Diet and exercise: effects on serum lipids, *Circulation,* 49(Suppl. 3, Abstr.), III-175,1974.
50. **Keen, H., Jarrett, R. J., and Alberti, K. G. M. M.,** Diabetes mellitus: a new look at diagnostic criteria, *Diabetologia,* 16, 283, 1979.
51. National Diabetes Data Group, Classification and diagnosis of diabetes mellitus and other categories of glucose intolerance, *Diabetes,* 28, 1039, 1979.
52. **Fuller, J. H., Shipley, M. J., Rose, G., Jarrett, R. J., and Keen, H.** Coronary-heart-disease risk and impaired glucose tolerance. The Whitehall Study, *Lancet,* 1, 1373, 1980.
53. Editorial, Impaired glucose tolerance and diabetes—WHO criteria, *Br. Med. J.,* 281, 1512, 1980.
54. **Stamler, R., Stamler, J., Dyer, A., Cooper, R., Collette, P., Berkson, D. M., Lindberg, H. A., Stevens, E., Schoenberger, J. A., Shekelle, R. B., Paul, O., Lepper, M., Garside, D., Tokich, T., and Hoeksema, R.,** Asymptomatic hyperglycemia and cardiovascular diseases in three Chicago epidemiologic studies, *Diabetes Care,* 2, 142, 1979.
55. **Garcia, M. J., McNamara, P. M., Gordon, T., and Kannel, W. B.,** Morbidity and mortality in diabetics in the Framingham population, *Diabetes,* 23, 105, 1974.
56. **Stout, R. W.,** Diabetes and atherosclerosis—the role of insulin, *Diabetologia,* 16, 141, 1979.
57. **Ducimetiere, P., Echwege, E., Papoz, L., Richard, J. L., Claude, J. R., and Rosselin, G.,** Relationship of plasma insulin levels to the incidence of myocardial infarction and coronary heart disease mortality in a middle-aged population, *Diabetologia,* 19, 205, 1980.
58. **Pyörälä, K.,** Relationship of glucose tolerance and plasma insulin to the incidence of coronary heart disease results from two population studies in Finland, *Diabetes Care,* 2, 131, 1979.
59. **Witztum, J. and Schonfeld, G.,** High density lipoproteins, *Diabetes,* 26, 326, 1979.
60. **Durrington, P. N.,** Serum high density lipoprotein cholesterol in diabetes mellitus: an analysis of factors which influence its concentration, *Clin. Chim. Acta,* 104, 11, 1980.
61. **Kennedy, A. L., Lappin, T. R. J., Lavery, T. D., Hadden, D. R., Weaver, J. A., and Montgomery, D. A. D.,** Relation of high density lipoprotein cholesterol concentration to type of diabetes and its control, *Br. Med. J.,* 2, 1191, 1978.
62. **Mann, J. I., Hoghson, W. G., Holman, R. R., Honour, A. J., Thorgood, M., Smith, A., and Baum, J. D.,** Serum lipids in treated children and their families, *Clin. Endocrinol.,* 8, 27, 1978.
63. **Sosenko, J. M., Breslow, J. L., Miettinen, O. S., and Gabbay, K. H.,** Hyperglycemia and plasma lipid levels. A prospective study of young insulin-dependent diabetic patients, *N. Engl. J. Med.,* 302, 650, 1980.
64. **Chase, H. P. and Glasgow, A. M.,** Juvenile diabetes mellitus and serum lipids and lipoprotein levels, *Am. J. Dis. Child.,* 130, 1113, 1976.
65. **Harno, K., Nikkilä, E. A., and Kuusi, T.,** Plasma HDL-cholesterol and postheparin plasma hepatic endothelial lipase (HL) activity: relationship to obesity and non-insulin dependent diabetes (NIDDM), *Diabetologia,* 19(Abstr.), 281, 1980.

66. **Mattock, M. B., Fuller, J. H., Maude, P. S., and Keen, H.,** Lipoproteins and plasma cholesterol in normal and diabetic subjects, *Atherosclerosis,* 34, 437, 1979.
67. **Nikkilä, E. A. and Hormila, P.,** Coronary heart disease and its risk factors among chronic insulin dependent diabetes, *Diabetologia,* 12(Abstr.), 412, 1976.
68. **Mattock, M. B., Salter, A., Fuller, J. H., and Omer, T.,** High density lipoprotein subfractions in insulin dependent diabetics, *Diabetologia,* 19(Abstr.), 298, 1980.
69. **Taskinen, M.-R. and Nikkilä, E. A.,** Lipoprotein lipase activity of adipose tissue and skeletal muscle in insulin-deficient human diabetes, *Diabetologia,* 17, 351, 1979.
70. **Reckless, J. P. D., Betteridge, D. J., Wu, P., Payne, B., and Galton, D. J.,** High-density and low-density lipoproteins and prevalence of vascular disease in diabetes mellitus, *Br. Med. J.,* 1, 883, 1978.
71. **Lopes-Virella, M. F. L., Stone, P. G., and Colwell, J. A.,** Serum high density lipoprotein in diabetic patients, *Diabetologia,* 13, 285, 1977.
72. **Elkeles, R. S., Wu, J., and Hambley, J.,** Haemoglobin A_1, blood glucose, and high density lipoprotein cholesterol in insulin-requiring diabetics, *Lancet,* 2, 547, 1978.
73. **Durrington, P.,** H.D.L. cholesterol in diabetes mellitus. Letter, *Lancet,* 2, 206, 1978.
74. **Ballantyne, D., White, C., Strevens, E. A., Lawrie, T. D. V., Lorimer, A. R., Manderson, W. G., and Morgan, H. G.,** Lipoprotein concentration in untreated adult onset diabetes mellitus and the relationship of the fasting plasma triglyceride concentration to insulin secretion, *Clin. Chim. Acta,* 78, 323, 1977.
75. **Carlson, L. A., Böttiger, L. E.,** Ischemic heart disease in relation to fasting values of plasma triglycerides and cholesterol. The Stockholm Prospective Study, *Lancet,* 1, 865, 1972.
76. **Hulley, S. B., Rosenman, R. H., Bawol, R. D., and Brand, R. J.,** Epidemiology as guide to clinical decisions. The association between triglyceride and coronary heart disease, *N. Engl. J. Med.,* 302, 1389, 1980.
77. **Nikkilä, E. A., Huttunen, J. K., and Ehnholm, C.,** Postheparin plasma lipoprotein lipase in diabetes mellitus. Relationship to plasma triglyceride metabolism, *Diabetes,* 26, 11, 1977.
78. **Wahren, J.,** Glucose turnover during exercise in healthy man and in patients with diabetes mellitus, *Diabetes,* 28(Suppl. 1), 82, 1979.
79. **Wahren, J., Felig, P., and Hagenfeldt, L.,** Physical exercise and fuel homeostasis in diabetes mellitus, *Diabetologia,* 14, 213, 1978.
80. **Wahren, J., Hagenfeldt, L., and Felig, P.,** Glucose and free fatty acid utilization in exercise, *Isr. J. Med. Sci.,* 11, 551, 1975.
81. **Berger, M., Berchtold, P., Cüppers, H. J., Drost, H., Kley, H. K., Müller, W. A., Wiegelmann, W., Zimmermann-Telschow, H., Gries, F. A., Krüskemper, H. L., and Zimmermann, H.,** Metabolic and hormonal effects of muscular exercise in juvenile type diabeties, *Diabetologia,* 13, 355, 1977.
82. **Koivisto, V. A. and Felig, P.,** Effects of leg exercise on insulin absorption in diabetic patients, *N. Engl. J. Med.,* 298, 79, 1978.
83. **Zinman, B., Murray, F. T., Vranic, M., Albisser, A. M., Leibel, B. S., McClean, P. A., and Marliss, E. B.,** Glucoregulation during moderate exercise in insulin treated diabetics, *J. Clin. Endocrinol. Metab.,* 45, 641, 1977.
84. **Kemmer, F. W., Berchtold, P., Berger, M., Starke, A., Cüppers, H. J., Gries, F. A., and Zimmermann, H.,** Exercise-induced fall of blood glucose in insulin treated diabetics unrelated to alteration of insulin mobilization, *Diabetes,* 28, 1131, 1979.
85. **Hagenfeldt, L.,** Metabolism of free fatty acids and ketone bodies during exercise in normal and diabetic man, *Diabetes,* 28(Suppl. 1), 66, 1979.
86. **Hagan, R. D., Marks, J. F., and Warren, P. A.,** Physiologic responses of juvenile-onset diabetic boys to muscular work, *Diabetes,* 28, 1114, 1979.
87. **Ruderman, N. B., Ganda, O. P., and Johansen, K.,** The effect of physical training on glucose tolerance and plasma lipids in maturity onset diabetes, *Diabetes,* 28(Suppl. 1), 89, 1979.
88. **Morris, J. N., Chave, S. P. W., Adam, C., Sirey, C., Epstein, L., and Sheehan, D. J.,** Vigorous exercise in leisure time and the incidence of coronary heart disease, *Lancet,* 1, 333, 1973.
89. **Morris, J. N., Everitt, M. G., Pollard, R., Chave, S. P. W., and Semmence, A. M.,** Vigorous exercise in leisure time: protection against coronary heart disease, *Lancet,* 2, 1207, 1980.
90. **Paffenbarger, R. S., Jr. and Hale, W. E.,** Work activity and coronary heart mortality, *N. Engl. J. Med.,* 292, 545, 1973.
91. **Paffenbarger, R. S., Jr., Wing, A. L., and Hyde, R. T.,** Physical activity as an index of heart attack risk in college alumni, *Am. J. Epidemiol.,* 108, 161, 1978.

Chapter 15

ANTIHYPERTENSIVE MEDICATION AND PHYSICAL TRAINING

Matti Uusitupa

TABLE OF CONTENTS

I. INTRODUCTION

There is strong evidence that physical training has beneficial influences on risk factors of coronary heart disease.[1] Physical exercise is also recommended as a part of management of mild hypertension, either combined with other types of nonmedical therapy or with antihypertensive drug treatment. In addition, some hypertensive patients treated successfully with antihypertensive drugs are physically active in their work or want to exercise in sports. For such patients the maintainence of physical capacity is of importance, and antihypertensive drugs which may limit their physical capacity may decrease patient compliance to antihypertensive therapy. These factors support the importance of knowing the effects of antihypertensive drugs on physical performance.

The present review will deal with effects of antihypertensive drugs on physical performance. Special attention is focused on the effects of β-adrenoceptor blocking agents because they are at present widely prescribed, especially for young hypertensive patients who are often physically active, too. Effects of diuretics and β-adrenoceptor blocking agents on plasma lipids and lipoproteins are also discussed.

II. PHYSICAL EXERCISE IN THE TREATMENT OF HYPERTENSION

Although some studies exist about beneficial effects of physical training on blood pressure levels in normotensive subjects,[2,3] in a controlled study 18 months' physical training did not lower the blood pressure levels in sedentary middle-aged men when compared to control subjects who remained physically inactive during the follow-up period.[4] On the contrary, in borderline hypertension[5] which is commonly associated with high cardiac output, rapid heart rate, and increased oxygen consumption at rest,[6–9] physical training may lower elevated blood pressure.[2,10,11] The effect of physical training in borderline hypertensive subjects reflects, in addition to the resting blood pressure values, upon the blood pressure response to exercise. One should be reminded, however, that the studies on the physical training in the treatment of borderline hypertension have been performed without control groups. The mode of action of physical training in borderline hypertension involves a normalization of the hyperkinetic circulation, and also a more economic myocardial work and oxygen consumption.[10,11] In addition, it has been shown that long-term physical training lowered elevated blood pressure in obese female subjects with hyperinsulism, and the decrease in the blood pressure was correlated with the initial concentrations and decreases in plasma insulin, triglycerides, and blood glucose.[12]

In established hypertension, the beneficial effect of physical exercise on blood pressure levels has been suggested to be negligible.[11] Promising results have been, however, reported in a group of hypertensive middle-aged men who were treated with antihypertensive drugs at the beginning of the study.[13] This study was, however, lacking in controls, too.

Although the efficacy of physical training in the treatment of mild hypertension remains so far unproven, the benefit of physical exercise, especially when combined with other nonmedical therapy of mild hypertension, such as salt restriction, weight reduction, and avoiding of inappropriate stress, can not be completely neglected. However, additional controlled studies concerning this topic are urgently needed. One should keep in mind that physical activity has other beneficial influences which may be of importance in primary prevention of ischemic heart disease.[14] In addition, physical exercise has been recommended as management of stress which may have a role in the pathogenesis of hypertension,[15] As regards the quality of physical training in the management of mild hypertension, dynamic exercise has been recommended, whereas static exercise may be deleterious because it may cause a marked increase in the blood pressure levels in hypertensive patients.[16]

III. EFFECT OF ANTIHYPERTENSIVE AGENTS ON PHYSICAL PERFORMANCE

A. β-Blocking Agents

It has been well documented that β-adrenoceptor blocking agents (β-blocking agents) increase working capacity of patients with angina pectoris. In addition to coronary heart disease, the use of β-blocking agents is becoming common as a primary treatment of mild hypertension. Although well tolerated, prescribing β-blocking agents to physically active persons may, however, involve some side effects which may cause a reduction in the physical fitness of these patients due both to the hemodynamic and metabolic consequences of β-blockade.

The β-blockade abolishes the exercise-induced effects of adrenergic stimulation, such as an increase in the pulse rate and blood pressure. In addition, β-blockade may decrease left ventricular contractility. Cardiac output decreases during exercise in patients treated with β-blocking agents, although stroke volume may slightly increase. An increase in the difference of arteriovenous oxygen saturation may partly compensate decreased cardiac output.[17–19] There is, however, no doubt that hemodynamic changes induced by β-blocking agents may cause a decline in maximal aerobic capacity. During vigorous exercise the reduction of physical fitness is most remarkable because cardiac output is closely dependent on the pulse rate response. It is obvious that the reduction of physical fitness is dependent on the dose of β-blocking agents used and thus on the degree of β-blockade.[20] Beta-blocking agents increase peripheral resistance of blood flow at the beginning of the treatment but during the long-term treatment and exercise this effect on blood flow may disappear.[17–19] $Beta_1$-selective β-blocking agents and those with intrinsic sympathomimetic activity may have less influence on peripheral circulation than nonselective β-blocking agents. In addition, β-blocking agents with intrinsic sympathomimetic activity may not diminish so much left vetricular contractility as do those agents without this property.[21,22] However, pindolol, a nonselective β-blocking agent with intrinsic sympathomimetic activity seems also to reduce maximal 0_2-uptake.[23]

In addition to hemodynamic influences of β-blocking agents, it has been suggested that metabolic consequences of β-blockade may contribute to the limitation of physical fitness observed in the treatment with β-blocking agents. Anderson et al.[24] and Fellenius et al.[25] have shown that both a $beta_1$-selective metoprolol and a nonselective β-blocking agent propranolol reduced maximal work performance and shortened the time until exhaustion. Similar results were obtained by Pearson et al.[26] who showed that both types of β-blocking agents decreased physical endurance assessed as either total work done or maximal work achieved. In addition, perceived exertion[27] during exercise was increased by β-blocking agents.

Reduction in physical fitness caused by β-blockade may be due to diminished muscle glycogenolysis,[23] or to the lack of free fatty acids[24–25] which are an important source of energy during physical exercise. An exercise-induced increase in serum potassium levels during treatment with β-blocking agents has been suggested to cause muscle fatigue which is commonly associated with the use of β-blocking agents.[28]

We have recently performed a placebo controlled study with double blind cross-over method in seven healthy male volunteers who participated in three bicycle ergometer tests, each lasting for 30 min, on the treatment of propranolol and metoprolol used in equipotent doses.[29] Both propranolol and metoprolol abolished the exercise-induced increase in the pulse rate and blood pressure similarly. Perceived exertion tended to increase during exercise during the treatment with both active agents. Two of the subjects stopped the exercise test when treated with metoprolol because of exhaustion. The most remarkable finding in metabolic consequences was an inhibition of the exercise-induced increase in plasma free fatty acids during the treatment with both propranolol and metoprolol. Muscle glycogenolysis was slightly decreased by β-blockade as assessed by muscle glycogen concentration before and after exercise, but the increase in blood lactate concentration during exercise was not influenced significantly by β-blocking agents. No significant differences were observed in serum potassium levels between the three experiments. Thus our results support the assumption that impairment of physical fitness and muscle fatigue caused by β-blockade may be explained by a deficiency of energy supply to exercising muscles due to the lack of free fatty acids.[24,25]

Nonselective β-blocking agents may cause a decline in blood glucose concentration during exercise,[23,30,31] but this effect does not seem to correlate with the reduction in physical performance.[29]

The effects of β-blocking agents on physical performance continues to attract increasing attention. Based on preliminary results it has been suggested that β-adrenergic blockade may prevent both exercise conditioning[32] and beneficial effects of aerobic exercise on blood lipids.[33]

B. Other Antihypertensive Agents

In addition to β-blocking agents, reserpin, diuretics, alpha-methyldopa, clonidine, and prazosin are useful antihypertensive drugs in the treatment of mild hypertension, whereas hydralazine should not be used alone because of reflex tachycardia and an increase in cardiac output.

Reserpin does not markedly influence the physical performance, but its use has declined in the last few years, particularly in Scandinavia, at least partly due to its side effects. The antihypertensive effect of diuretics has been documented both at rest and during exercise.[34,35] Diuretics cause a decrease in plasma volume also during long-term treatment. They reduce total peripheral resistance but do not affect heart rate, stroke volume, or cardiac output. The hemodynamic influences of diuretics do not impede physical performance, hypokalemia induced by diuretics may, however, do so. Alpha-methyldopa has been shown to cause a slight reduction in cardiac output and in total peripheral resistance.[36] Its antihypertensive effect is, however, small during vigorous exercise. Physical fitness remains fairly unaffected during the treatment with alpha-methyldopa, but orthostatic hypotension and other side-effects limit its usefulness.

Long-term treatment with clonidine has been shown to cause a decrease in the heart rate and total peripheral resistance. A slight decline in cardiac output has been observed at rest, but exercise-induced increase in cardiac output is preserved during clonidine treatment.[37,38] The most common side effects of clonidine are drowsiness and dryness of the mouth. Although these side effects tend to decrease during long-term therapy, the drug remains unacceptable for many patients.

The hemodynamic influences of prazosin do not lead to a reduction of physical fitness. At rest a decrease in blood pressure is associated with a decrease in peripheral resistance.[38,39] Cardiac output remains unaffected. During exercise, on the other hand, cardiac output increases due to the increase in stroke volume. Prazosin seems to be suitable for patients who practice sports. Furthermore, it has no unfavorable effects on plasma lipids.[40]

IV. ANTIHYPERTENSIVE AGENTS AND PLASMA LIPIDS

There occurs now a great deal of evidence that both diuretics and β-blocking agents may have untoward effects on plasma lipids. As early as 1964 thiazide diuretics were reported to increase serum cholesterol levels.[41] Later it was found that, in addition to serum cholesterol, serum triglycerides increased in patients treated with thiazide diuretics.[42] These as well as many other subsequent clinical observations have been confirmed by a recent controlled study[43] which showed that both hydrochlorothiazide and chlorthalidone increased plasma total cholesterol and triglycerides. A significant increase was observed in very low density lipoprotein (VLDL)-cholesterol during both drugs. Low density lipoprotein (LDL)-cholesterol increased also, but the change was significant only during chlorthalidone treatment. No consistent changes were observed in high density lipoprotein (HDL)-cholesterol as compared to placebo treatment. A cholesterol lowering diet could largely prevent the changes in plasma lipids. According to some studies, diuretics have decreased plasma HDL-cholesterol too.[44,45] Furosemide, distinct from thiazide diuretics, does not affect LDL-cholesterol.[46]

As regards β-blocking agents, they may induce a significant increase in serum triglycerides and a decrease in HDL-cholesterol without affecting total cholesterol concentration.[40,47–49] In a study concerning long-term effects of sotalol therapy on plasma lipids a significant increase in plasma total cholesterol was also found.[50]

At present, data suggest that all β-blocking agents affect plasma lipids similarly. The mechanism of action on plasma lipids of diuretics or of β-blocking agents remains unknown. It is, however, no doubt that both of these popular antihypertensive drugs may cause a significant reduction in HDL/LDL + VLDL cholesterol ratio and thus they may potentially increase atherogenesis and hamper the benefit associated with the blood pressure lowering effect. Whether or not this cancellation in reality occurs, has not yet been shown in longitudinal follow-up studies. The potentially untoward effects of antihypertensive agents give, however, a good indication to study nonpharmaceutical methods for the therapy of mild hypertension.

V. SUMMARY AND CONCLUSIONS

Beta-blocking agents are at present commonly used in the treatment of mild hypertension. Although these agents are lacking serious side effects, they may cause a reduction in physical fitness. The impairment in physical fitness and muscle fatigue observed during the treatment with β-blocking agents may be explained by a deficiency of energy supply due to the lack of free fatty acids. In addition, a decrease in cardiac output induced by β-blockade may result in a reduction in physical performance. Other antihypertensive agents used in mild hypertension, diuretics, and prazosin, are in general well tolerated by physically active patients.

Both diuretics and β-blocking agents reduce HDL/LDL + VLDL cholesterol ratio and thus potentially increase the risk of atherogenesis. Whether or not this untoward effect hampers the benefit of antihypertensive effect of these drugs has not been documented.

REFERENCES

1. **Paffenbarger, R. S., Jr. and Hyde, R. T.,** Exercise as protection against heart attack, *N. Engl. J. Med.*, 302, 1026, 1980.
2. **Choquette, G. and Ferguson, R. J.,** Blood pressure reduction in ''borderline'' hypertensives following physical training, *Can. Med. Assoc. J.*, 108, 699, 1973.
3. **Bonnano, J. A. and Lies, J. E.,** Effects of physical training on cornary risk factors, *Am. J. Cardiol.*, 33, 760, 1974.

4. **Pyörälä, K., Käräva, R., Punsar, S., Oja, P., Teräslinna, P., Partanen, T., Jääskeläinen, M., Pekkarinen, M.-L., and Koskela, A. A.,** Controlled study of the effects of 18 months' physical training in sedentary middle-aged men with high indexes of risk relative to coronary heart disease, in *Coronary Heart Disease and Physical Fitness,* Munksgaard, Copenhagen, 1971, 261.
5. **Julius, S., Hansson, L., Andrén, L., Gudbrandsson, T., Sivertsson, R., and Svensson, A.,** Borderline hypertension, *Acta Med. Scand.,* 208, 481, 1980.
6. **Sannerstedt, R.,** Haemodynamic responses to exercise in patients with arterial hypertension, *Acta Med. Scand. Suppl.,* 458, 45, 1966.
7. **Lund-Johansen, P.,** Haemodynamics in early essential hypertension, *Acta Med. Scand.,* 482, 9, 1967.
8. **Julius, S. and Conway, J.,** Haemodynamic studies in patients with borderline blood pressure elevation, *Circulation,* 38, 382, 1968.
9. **Lund-Johansen, P.,** Central haemodynamics in essential hypertension, *Acta Med. Scand. Suppl.,* 606, 35, 1977.
10. **Hanson, J. and Nedde, W. H.,** Preliminary observations on physical training for hypertensive males, *Circulation Res.,* 27(Suppl. 1), 49, 1970.
11. **Sannerstedt, R., Wasir, H., Henning, R., and Werkö, L.,** Systemic haemodynamics in mild arterial hypertension before and after physical training, *Clin. Sci. Mol. Med.,* 45, 145, 1973.
12. **Krotkiewski, M., Mandroukas, K., Sjöström, L., Sullivan, L., Wetterqvist, H., and Björntorp, P.,** Effects of long-term physical training on body fat, metabolism, and blood pressure in obesity, *Metabolism,* 28, 650, 1979.
13. **Boyer, J. L. and Kasch, F. W.,** Exercise therapy in hypertensive men, *JAMA,* 211, 1668, 1970.
14. **Pyörälä, K.,** Life habits (Nutrition, non-smoking and physical activity) in the primary prevention of ischaemic heart disease, in *Int. Congr. Ser. No. 470,* Proc. 8th World Congr. Cardiology, Hayase, S. and Murao, S., Eds., Exerpta Medica, Amsterdam, 1979, 120.
15. **Eliot, R. S.,** Stress and cardiovascular disease, *Eur. J. Cardiol.,* 5, 97, 1977.
16. **Ewing, D. J., Irving, J. B., Kerr, F., and Kirby, B. J.,** Static exercise in untreated systemic hypertension, *Br. Heart J.,* 35, 413, 1973.
17. **Lund-Johansen, P. and Ohm, O. J.,** Haemodynamic long-term effects of β-receptor-blocking agents in hypertension: a comparison between alprenolol, atenolol, metoprolol and timolol, *Clin. Sci. Mol. Med.,* 51, 481, 1976.
18. **Adolfsson, L. and Sonta, C.,** Haemodynamic effects of two cardioselective beta-adrenoceptive antagonists, metoprolol and H 87/07 in coronary insufficiency, *Scand. J. Clin. Lab. Invest.,* 36, 755, 1976.
19. **Frisk-Holmberg, M., Jorfeldt, L., and Juhlin-Dannfelt, A.,** Influence of alprenolol on haemodynamic and metabolic responses to prolonged exercise in subjects with hypertension, *Clin. Pharmacol. Ther.,* 21, 675, 1977.
20. **Folgering, H. and van Bussel, M.,** Maximal exercise power after a single dose of metoprolol and of slow-release metoprolol, *Eur. J. Clin. Pharmacol.,* 18, 225, 1980.
21. **Lysbo Svendsen, T., Hartling, O., and Trap-Jensen, J.,** Immediate haemodynamic effects of propranolol, practolol, pindolol, atenolol and ICI 89,406 in healthy volunteers, *Eur. J. Clin. Pharmacol.,* 15, 223, 1979.
22. **Franciosa, J. A., Johnson, S. M., and Tobian, L. J.,** Exercise performance in mildly hypertensive patients, impairment by propranolol but not oxprenolol, *Chest,* 78, 291, 1980.
23. **Franz, I.-W. and Lohman, F. W.,** The influence of long-term cardioselective and non-selective β-receptorblockade on blood pressure, 0_2-uptake and cardohydrate metabolism, *Z. Kardiol.,* 68, 503, 1979.
24. **Anderson, S. D., Bye, P. T. P., Perry, C. P., Hamor, G. P., Theobald, G., and Nyberg, G.,** Limitation of work performance in normal adult males in the presence of beta-adrenergic blockade, *Aust. N. Z. J. Med.,* 9, 515, 1979.
25. **Fellenius, E., Carlsson, E., Åström, M., Lipska, M., Lundborg, P., Svensson, L., Bengtsson, C., and Smith, U.,** Metabolic Effects of β-Adrenoceptor blockade during Exercise, in *4th Int. Symp. Biochemistry of Exercise,* (Abstr.), Brussels, June 18 to 21, 1979, 12.
26. **Pearson, S. P., Banks, D. C., and Patrick, J. M.,** The effect of β-adrenoceptor blockade on factors affecting exercise tolerance in normal man, *Br. J. Clin. Pharmacol.,* 8, 143, 1979.
27. **Borg, G.,** Perceived exertion as an indicator of somatic stress, *Scand. J. Rehab. Med.,* 2, 92, 1970.
28. **Carlsson, E., Fellenius, E., Lundborg, P., and Svensson, L.,** β-adrenoceptor blockers, plasma potassium and exercise, *Lancet,* 2, 424, 1978.
29. **Uusitupa, M., Siitonen, O., Härkönen, M., Gordin, A., Aro, A., Hersio, K., Johansson, G., Korhonen, T., and Rauramaa, R.,** Metabolic and hormonal response to physical exercise during $beta_1$-selective and non-selective beta-blockade, submitted.
30. **Bewsher, P. D.,** Propranolol, blood sugar, and exercise, *Lancet,* 1, 104, 1967.

31. **Linton, S. P., Scott, P. H., and Kendall, M. J.,** Blood sugar and beta-blockers, *Br. Med. J.*, 2, 877, 1976.
32. **Sable, D. L., Brammell, H. L., Sheehan, M. W., Nies, A. S., and Horwitz, L. D.,** Beta-adrenergic blockade prevents exercise conditioning, *Circulation*, 62(Suppl. 3), 201, 1980.
33. **Brammell, H. L., Sable, D. L., Horwitz, L. D.,** Beta-adrenergic blockade prevents effects of aerobic conditioning on blood lipids, *Circulation*, 62(Suppl. 3), 122, 1980.
34. **Conway, J. and Lauwers, P.,** Haemodynamic and hypotensive effects of long-term therapy with cholorothiazide, *Circulation*, 21, 21, 1960.
35. **Lund-Johansen, P.,** Hemodynamic changes in long-term diuretic therapy of essential hypertension, *Acta Med. Scand.*, 187, 509, 1970.
36. **Lund-Johansen, P.,** Hemodynamic changes in long-term alpha-methyldopa therapy of essential hypertension, *Acta Med. Scand.*, 192, 221, 1972.
37. **Lund-Johansen, P.,** Hemodynamic effects of clonidine in man, in *Regulation of Blood Pressure by the Central Nervous System*, Onesti, G., FErnandes, M., and Kim, K. E., Eds., Grune & Stratton, New York, 1976, 355.
38. **Onesti, G. and Fernandes, M.,** Recent acquisitions in anithypertensive therapy: clonidine, minoxidil and prazosin, in *Hypertension: Mechanisms, Diagnosis and Treatment*, Cardiovascular Clinics Ser., Vol. 9(1), Onesti, G. and Brest, A. N., Eds., F. A. Davis, Philadelphia, 1978, 273.
39. **Lund-Johansen, P.,** Hemodynamic changes at rest and during exercise in longterm prazosin therapy of essential hypertension, in *Prazosin-Evaluation of a New Antihypertensive Agent*, Cotton, D. W. K., Eds., Excerpta Medica, Amsterdam, 1974, 64.
40. **Leren, P., Helgelnad, A., Holme, I., Foss, P. O., Hjerman, I., and Lund-Larsen, P. G.,** Effect of propranolol and prazosin on blood lipids. The Oslo study, *Lancet*, 2, 4, 1980.
41. **Schoenfield, M. R. and Goldberger, E.,** Hypercholesterolemia induced by thiazides: a pilot study, *Curr. Ther. Res.*, 6, 180, 1964.
42. **Ames, R. P. and Hill, P.,** Increase in serum-lipids during treatment of hypertension with chlorthalidone, *Lancet*, 1, 721, 1976.
43. **Grimm, R. H., Jr., Leon, A. S., Hunninghake, D. B., Lenz, K., Hannan, P., and Blackburn, H.,** Effects of thiazide diuretics on plasma lipids and lipoproteins in mildly hypertensive patients, *Ann. Int. Med.*, 94, 7, 1981.
44. **Bauer, J. H., Brooks, C. S., Weinstein, I., Wilcoz, H., and Heimberg, M.,** Comparison of ticrynafen and hydrochlorotiazide on plasma lipids, *Circulation*, 59, 11, 1979.
45. **Helgelnad, A., Hjermann, I., Leren, P., Enger, S., and Holme, I.,** High density lipoprotein cholesterol and antihypertensive drugs, *Br. Med. J.*, 2, 403, 1978.
46. **Joos, C., Kewitz, H., and Reinhold-Kourniati, D.,** Effects of diuretics on plasma lipoproteins in healthy men, *Eur. J. Pharmacol.*, 17, 251, 1980.
47. **Tanaka, N., Sakaguchi, S., Oshige, K., Niimura, T., and Kanehisa, T.,** Effect of chronic administration of propranolol on lipoprotein composition, *Metabolism*, 25, 1071, 1976.
48. **England, J. D. F., Hua, A. S. P., and Shaw, J.,** β-Adrenoreceptor-blocking agents and lipid metabolism, *Clin. Sci. Mol. Med.*, 55, 323, 1978.
49. **Day, J. L., Simpson, N., Metcalfe, J., and Page, R. L.,** Metabolic consequences of atenolol and propranolol in treatment of essential hypertension, *Br. Med. J.*, 1, 77, 1979.
50. **Lehtonen, A. and Viikari, J.,** Long-term effect of sotalol on plasma lipids, *Clin. Sci.*, 57, 405, 1979.

Chapter 16

PROGNOSTIC AND PREVENTIVE VALUE OF EXERCISE IN CORONARY HEART DISEASE

Eino Hietanen

TABLE OF CONTENTS

I. EPIDEMIOLOGICAL STUDIES

In attempts to evaluate the use and promptness of exercise programs to persons having coronary heart disease (CHD), the first task is to analyze the present knowledge of the metabolic and physiological effects of exercise training on cardiorespiratory and other body functions. When judging exercise programs and motivating people to exercise, one should consider various aspects before realizing the exercise program. No definite indications exist that the exercise training decreases mortality on cardiovascular diseases or prevents or slows the progress of coronary heart disease! However, quite strong indirect evidence exists favoring the prognostic significance of physical activity in preventing coronary heart disease both from retrospective and prospective epidemiological studies (Table 1). Most of these studies have dealt with the correlation between physical activity at work or during leisure time and the mortality in cardiovascular diseases (myocardial infarction). Most of the retrospective studies have shown a decreased incidence of CHD or decreased mortality in CHD in physically active men (Table 1). There are, however, also studies where no such correlation could be found, although even these studies did not show any disadvantageous effects of exercise either (Table 1A).

All prevalent or cross-sectional studies favor the positive effect of physical activity in the prevention of the CHD (Table 1B). At present, most valid epidemiological studies are possibly prospective follow-ups of human groups where the incidence of CHD is followed in the long run. Most of these studies have shown that high physical activity at work or during leisure time prevents CHD (Table 1C), although results exist where no effects or no relationship were established (Table 1C). Paffenbarger et al.[35] found that heavy physical activity at work decreased the mortality in CHD independently from the effects of smoking, blood pressure, or weight. Morris and co-workers[27] have studied the significance of leisure time physical activity on the progression of CHD. They found in those, who were active, a much lower incidence of CHD than in those inactive during their leisure time. Pathological autopsies have not revealed any good correlation between the life-time exercise habits and the condition of coronary arteries (Table 1D). Possibly, the most favorable studies in terms of physical exercise preventing CHD or promoting recovery from the myocardial infarction are the rehabilitation studies (Table 1E). Only in one study was no effect found when some of the recent studies were collected.

II. LIPOPROTEINS AND CORONARY ARTERY OBLITERATION

Numerous studies have confirmed the role of high density lipoproteins (HDL) as an independent risk factor for CHD separate from other lipids.[55,56] It has been calculated that this lipoprotein fraction might be one of the strongest lipid parameters to predict CHD in the way that the lower the HDL concentration the higher the risk to have CHD.[57] In myocardial infarction patients, HDL cholesterol concentration is lower than in controls,[58–62] and also in patients with CHD the HDL cholesterol concentration is low.[55,63,64] Recently, the apoprotein determinations have confirmed the data obtained with lipid determinations.[65] In patients with CHD these apoproteins typically present in LDL and VLDL fractions were elevated, i.e., apoproteins B and C, and the apoprotein present in HDL fraction, apo A, was decreased.[65]

Lately, many correlation studies have been made between the angiographic image of coronary arteries and the HDL cholesterol levels. These studies have quite convincingly demonstrated that the more obliterated coronaries the lower the HDL cholesterol concentration in plasma.[57,66–68] Pearson et al.[67] correlated the degree of coronary artery

Table 1
RELATIONSHIP OF PHYSICAL ACTIVITY TO CORONARY HEART DISEASE[2–54]

	Physical activity vs. decrease in CHD or CHD mortality
A. Retrospective epidemiologic studies	
England and Wales mortality statistics, 1963	+
London busmen, 1953	+
Malmo, 1958	–
North Dakota, 1959	+
Chicago mortaility statistics, 1960	–
California mortality statistics, 1960	+
Canadian VA, 1961	–
U.S. railmen, 1962	+
U.S. postmen, 1963	+
South African railroad, 1963	–
Health Insurance Plan of New York, 1966	+
Israel kibbutzi, 1966	+
Harvard alumni	+
Positive	9
Negative	4
B. Prevalence or cross-sectional studies	
People's Gas Company, 1960	+
U.S. railroad, 1962	+
Evans County, Ga., 1965	+
Stockholm-Edinburgh study, 1979	+
Positive data	4
C. Prospective studies	
Los Angeles civil servants, 1964	–
London busmen, 1966	+
Chicago Electric Company, 1969	–
U.S. railroad, 1970	–
People's Gas Company, Chicago, 1970	–
Seven country study, 1970	–
San Francisco longshoremen, 1970, 1975, 1977	+
Western collaborative group, 1970 and 1975	+
Gothenburg, Sweden, 1971, 1975, 1978	(+)/–
Framingham, Mass. 1971	+
Evans County, Ga., 1971	(+)
London civil service, 1973, 1978	+
National Coronary Drug Study, 1975	+

Table 1 (continued)
RELATIONSHIP OF PHYSICAL ACTIVITY TO CORONARY HEART DISEASE[2–54]

C. Prospective studies	Physical activity vs. decrease in CHD or CHD mortality
Puerto Rico Heart Study, 1978	
Urban Men	+
Rural Men	−
Positive data	9
Negative data	7
D. Autopsy studies	
England, 1958	(+)
Westchester, 1960	−
Oxford, 1967	(+)
Jaffa, 1970	−
Finland, 1975	−
Positive (weak)	2
Negative	3
E. Rehabilitation studies	
Tel Aviv, 1968	+
Cleveland, 1969	+
Rechnitzer et al., 1972	+
Wilhelmsen et al., 1975	−
Redwood et al., 1972	+
Kattus et al., 1972	+
Kavanagh and Shephard,	+
1974	+
Fletcher and Cantwell,	
1974	+
Ferguson et al., 1975	+
Kennedy et al., 1976	+
Positive	9
Negative	1

obstruction and the involvement of the coronary arteries with the plasma HDL cholesterol concentration both in men and women. They found decreased HDL cholesterol levels in those having angiographically diagnosed CHD. The involvement of the left main coronary artery caused a marked decrease of HDL cholesterol; the more arteries that were diseased the greater the decrease in HDL cholesterol. Although women have higher plasma HDL cholesterol concentration normally than men, in CHD the decreased HDL cholesterol concentrations are nearly equal. In persons of older age (50 to 70 years) low HDL cholesterol was also at least as good a predictor for angiographically diagnosed CHD as in younger ages.[67] In another study by Wieland et al.[68] the predictability of HDL cholesterol regarding the existence of CHD confirmed agiographically was worse than in the study by Pearson.[67] Also in this study by Wieland et al.,[68] the HDL/LDL (low density lipoprotein) ratio was a good determinant of the presence of CHD. Tan et al.[62] also studied serum lipids in persons after coronary angiography due to the evaluation of the coronary arteries because of chest pain. Some of the patients had had myocardial infarction. In this study no grading of coronary arteries was made except that abnormal was determined as over 50% stenosis in at least one artery. The data showed the lowest HDL cholesterol concentrations in those patients with abnormal coronary arteries and highest HDL levels in those of the control persons, independently of the sex.[62] As significant as the value of HDL cholesterol is the percentage of HDL cholesterol of the total plasma cholesterol level, also according to this study. The relative values of HDL cholesterol were lowest (17.7%) in the group with obliterated coronary artery(ies) and highest in the healthy persons (27.1%).[62]

Exercise training and physical activity increase HDL cholesterol concentration in plasma.[69–79] When high HDL cholesterol is related to the decreased CHD risk,[80,81] one might expect that training would prevent CHD by increasing HDL level as this lipoprotein possibly transports cholesterol from vascular walls to the liver for further metabolism.[82,83]

III. EXERCISE AND RISK FACTORS

Although no direct evidence on the positive effect of exercise in the prevention of CHD exists; the possible positive effects of exercise are supported by data that it may decrease known risk factors.[56,84–90] The risk factors include numerous factors—both metabolic, physiological, pharmacological, and personal lifestyle—which have been found to be related to CHD despite no cause and consequently proof has yet to be found (Table 2). The metabolic factors include high plasma total cholesterol and triglyceride levels and low HDL cholesterol. Moreover, these include diabetes and gout as well as disorders of lipid metabolism. Nonmetabolic risk factors, either directly or indirectly related to CHD, are manyfold. Well-known and generally acknowledged as risk factors are high blood pressure, cigarette smoking, personal type, and male sex.[84,85] Dietary factors may not be direct risk factors but they may have indirect effects by regulating, in part, blood lipids and development of obesity. It is known that the genetic factors have effects in the development of CHD, either directly or indirectly. The use of oral contraceptives in the presence of other risk factors may increase the possibility of having coronary heart disease.

Most studies where physical activity has been found to decrease risk factors of CHD have been conducted in healthy persons. From preceding chapters it has become evident that exercise training decreases triglyceride and total cholesterol concentrations in blood and increases HDL cholesterol.[55,72,73,75,76,79,92,93] Physical activity also possibly decreases blood pressure although discrepant data exist.[88,94–98]

Both the changes in blood lipids and blood pressure might, to a part, be mediated by changes in the body weight or composition caused by exercise. Often physical activity does not decrease weight but increases the relative proportion of the muscle tissue and decreases the proportion of the adipose tissue as judged from skin-fold thicknesses. Despite the fact that no definite positive effects of physical activity on all risk factors have been found, no risk factors have increased due to the exercise in any study.[95,96,98,99]

Physical activity also affects blood clotting mechanisms.[100] The response in blood clotting to exercise depends on the length of the training or exercise period.[100] Immediately after an exercise stress test and after a short-time physical performance no change in prothrombin time is found and thromboplastin time is shortened, and the amount of factor VIII is increased. In very fit persons the shortening clotting time after the exercise stress test is much less than in nonfit persons.[100]

The aggregation of platelets is an important step following the endothelial damage. Platelet function can be evaluated by their number, adhesion, and aggregation. Acute physical stress increases the number of platelets in the circulation.[100] The reason for the increased number of platelets after exercise is unknown but possibly the pulmonary capillary bed is a reserve for platelets and releases them along with increased circulation. The release of platelets is lower in physically active persons than in sedentary persons.[100] On the other hand, the platelet adhesion decreases due to a moderate activity such as walking. The platelet aggregation decreases in short-term exercise stress tests

Table 2
RISK FACTORS OF CORONARY HEART DISEASE

Metabolic
- High total cholesterol (mainly LDL cholesterol)
- Hypertriglyceridemia
- Low HDL cholesterol
- Diabetes
- Hyperlipoproteinemias
- Gout

Nonmetabolic
- Hypertension
- Cigarette smoking
- Nutritional factors
- Obesity
- Physical inactivity
- Genetic factors
- Male sex
- Age
- Oral contraceptives combined with other risk factors

both in healthy persons and in those having CHD.[100] The connection of the platelet aggregation and adhesion to physical fitness is partly unsolved but even a moderate or light training is known to decrease the aggregation. In the formation of blood clot the fibrinolytic activity also plays an important role. It is generally accepted that physical activity increases a fibrinolytic activity in blood.[99,100] As the intensity and duration of physical activity increases, the fibrinolytic activity also increases. The ratio of physical intensity to a person's maximal physical performance is a significant determinant of the fibrinolytic activity. In healthy persons the exercise at a level of 80 to 85% of the-maximal pulse rate also increases the fibrinolytic activity in acute exercise.[100]

The endothelial damage followed by a clot formation is one of the mechanisms thought to lead to the formation of an atheromatous plaque in CHD.[101] Atherosclerotic patients have been found to have a decreased fibrinolytic exercise response as also do diabetics and type IV hyperlipidemic persons.[100] Although the connection between blood clotting mechanisms and the physical activity is not completely solved, the fact remains that increased fibrinolytic activity and decreased platelet aggregation due to the physical exercise training are protective mechanisms in terms of the prevention of CHD.

Physical activity also modifies hemodynamics (Table 3).[95,99,103,104] No clear-cut increase in the collateral formation due to a physical activity has been found in man but experimental studies do support this hypothesis.[91,102] In some studies persons with CHD showed improved ECG response following improved physical fitness due to the training program.[96] The stroke volume increases due to the physical fitness both at rest and during exercise; this, however, precludes normal function of the heart without, e.g., dyskinesia due to the myocardial infarction. Physical activity improves, at least in healthy persons, maximal oxygen uptake, lowers resting heart rate and improves the load response of the heart rate.

Based on the research in healthy persons as well as in those with CHD it is evident that the physical activity might decrease the progress of CHD or prevent its development. Thus, it seems proper to motivate a person with CHD as well as healthy persons to practice exercise. However, before preceding to prescribe exercise to persons with CHD or other cardiac problems, certain precautions must be considered.

Table 3
EFFECT OF PHYSICAL TRAINING ON CARDIOVASCULAR, RESPIRATORY, AND METABOLIC FUNCTIONS

Parameter	Rest	Maximal work load
Heart rate	Decrease	Decrease or unchanged
Cardiac hypertrophy	Increase	—
Minute volume	Unchanged	Increase
Stroke volume	Increase	Increase
Hemoglobin and blood volume	Increased	—
Blood pressure		
Normotension	Unchanged or decrease	Unchanged
Hypertension	Decrease	Decrease
Lung volumes	Increase	—
Diffusion capacity	Increase?	Increase
Maximal voluntary ventilation	Unchanged	Increase
Oxygen uptake	Unchanged	Increase
Muscle oxygen utilization	Unchanged	Improved
Muscle glycogen depletion	—	Improved

Table 4
MEDICAL EXAMINATION OF PATIENTS WITH CORONARY HEART DISEASE PRECEDING EXERCISE PROGRAMMING

Physical examination
Blood tests
    Hemoglobin, sedimentation rate, white cell count, creatinine (or BUN)
Cardiological thorax X-ray
Spirometry
Exercise stress test (treadmill or ergometer)
Echocardiography

IV. PHYSICAL EXAMINATION PRECEDING EXERCISE TRAINING

Preceding exercise programs, persons with CHD or who are suspected to have CHD, as well as healthy persons over 40 years of age must have a careful medical examination to estimate the progressive exercise program as well as to detect possible contraindications for exercise (Table 4). A careful medical interview is essential to find out habits of individual physical activity to estimate the anticipated performance level. Attention must be paid to the physical strain at work and to the amount of physical exercise during leisure time. Symptoms of diseases such as chest pain must also be addressed. The use of therapeutic drugs is essential to know before exercise programming due to the possible adverse effects of drugs on physical performance. The use of β-blocking drugs is common among CHD patients and in these cases the exercise prescribing cannot be made exclusively on heart rate basis. In exercise stress tests, whether using treadmill or ergometer, the slow heart rate response to increases of work loads is typical when β-blocking drugs are used.

Attention must be paid to the auscultation of heart and lungs, possible cardiac insufficiency, blood pressure, and arrhythmias in the physical examination. The examination in the cases of persons with heart troubles or those over 40 years of age must include resting ECG and lung function test. Also of importance is the health of muscles and joints as well as peripheral arteries. Laboratory tests must be done to exclude anemias, and to cover the possible risk factors for CHD; also blood lipid analysis can be performed. The exercise stress test, either treadmill or ergometer test, gives a good estimation on the cardiovascular stress tolerance and helps to detect CHD and the exercise response of existing CHD. The most informative exercise test is to an individual maximum whether based on the age adjusted maximal heart rate or to the appearance of exercise test limiting symptoms. The exercise stress test is regarded as a good indicator to estimate the performance both after the myocardial infarction and in the coronary heart disease.[105–107] Whether the cardiac status is not clear on the basis of the stress test, the echocardiographic examination gives further information on the structure of the heart valves, dimensions of the heart, and also on the function of the heart.

Based on the cardiovascular and other examinations, the possible contraindications for exercise training are easily recognized (Table 5). The contraindications for the exercise training can be divided as metabolic, cardiovascular, and pulmonary contraindications or limitations. It is of importance not to miss other diseases a person with CHD might have. The cardiovascular contraindications for the exercise training include recent (less than 4 weeks) myocardial infarction, uncompensated cardiac insufficiency, aortic valve stenosis, uncontrolled arrhythmias, myocarditis, and obstructive cardiomyopathy. Limitations include multifocal ventricular arrhythmias and consecutive ventricular ectopic beats and developing cardiac insufficiency in the exercise stress test. Also, exercise-induced obstructive decrease in lung function tests is a limitation but not a contraindication for the exercise training. The use of chromoglygate in conjunction with exercise might prevent the obstructive lung symptoms.

Other diseases hampering or limiting the exercise training are labile diabetes, adrenal insufficiency, hyperthyroidism, and serious kidney and liver insufficiencies. Temporary limitations may be caused by acute infections and arthritis. However, many of the contraindications and limitations are such that they can be treated and the situation can be reestimated after adequate therapy. Typically temporary treatable limitations include cardiac insufficiency, arrhythmias, and acute infections. The use of sublingual nitroglycerin or nitroglycerin ointment can relieve chest pain symptoms and improve the physical performance in very symptom-limited persons.

V. PLANNING OF TRAINING PROGRAM IN CORONARY HEART DISEASE

The training program must be progressive both in patients with CHD and in healthy but sedentary persons. Thus, the goal is to improve the physical fitness gradually to the optimal maintenance level. Both the minute volume of the heart and the heart rate increase linearly with increase in oxygen uptake. Heart rate is the main determinant of the minute volume in load levels exceeding 50% of the individual maximal heart rate.[96] The following of the heart rate is the main indicator to estimate the increase in minute volume due to the increase in work load at exercise. Another determinant commonly used is the product of heart rate and systolic blood pressure.

The pulmonary ventilation is also linearly correlated with the relative load of the maximal working capacity up to the level of 85% of the maximal oxygen uptake at which level the anaerobic production of lactic acid stimulates respiration to a relatively larger extent.[96] Thus, at work loads under 80% of the maximal oxygen uptake, the increase in ventilation and the increase in heart rate are related to each other best.

In the process of planning an exercise program the individual maximal oxygen uptake or symptom limited maximal performance capacity forms the basis. Whether the maximal performance capacity is reached at exhaustion without other symptoms the exercise

Table 5
CONTRAINDICATIONS AND LIMITATIONS FOR EXERCISE TRAINING

Contraindications

Cardiovascular

Recent myocardial infarction
Uncompensated cardiac insufficiency
Uncontrolled arrhythmias
Labile angina pectoris
Second and third degree atrio-ventricular blocks
Transient ischemic attacks

Exercise stress test

Multifocal ventricular ectobic beats
Cardiac insufficiency in exercise
Angina pectoris at low work load and deep ST-segment descent (>0.4 mV)

Noncardiovascular

Difficult obstructive or restrictive lung disease
Cor pulmonale
Labile diabetes
Hyperthyroidism
Adrenal insufficiency
Acute infection
Thrombophlebitis
Active arthritis

Limitations

Uncontrolled hypertension
Stress asthma
Massive obesity
Unoperated disc (lumbo-sacral region) prolapse
Drugs interacting with hemodynamics

program can be planned so as to give a pulse rate of 80 to 85% of the maximal age-predicted heart rate. In the case the exercise stress test has been symptom-limited, the heart rate leading to the symptom causing the interruption of the stress test must be regarded as an individual maximal heart rate. In those persons having β-blocking drugs, a lower than 80 to 85% of the age-predicted maximal heart rate must also be the objective.

In the programming of the exercise training for persons with CHD a gradual increase in the work load is most acceptable.[108,109] In a cross-sectional study by Erkelens et al.[59] the physical activity levels of postinfarction persons were graded at 15 levels. When the starting level has been determined based on exercise stress test the grades are increased in accordance with the progress of training until a maintenance level is obtained. A study resembling this was conducted recently by Streja and Mymin[109] with persons who had either CHD, infarction, or who had had a by-pass operation. In this study the participants had a treadmill stress test to estimate the maximal performance capacity. The training consisted of indoor running, jogging, walking, and calisthenics. The exercise level was increased up to 70 to 85% of the maximal heart rate determined with the treadmill stress test and this heart rate was maintained 20 to 30 min three times a week for 3 months. During this time the distance jogged or walked remained as 2.8 km per performance but the speed increased from 6.25 km/hr to 7.01 km/hr. After the

3-month follow-up the treadmill test had improved significantly with the simultaneous increase in the HDL cholesterol concentration.[109]

Preceding the elevation of exercise level to the estimated heart rate there must be a warm-up period of 5 to 10 min to increase the muscular and circulatory adaptation to exercise and to minimize the symptoms like muscle pains and chest pain due to CHD. Also, the exercise period should be followed by 10 min of calisthenics or other light exercise to prevent the symptoms caused by a sudden cessation of the exercise.

When a person having CHD first starts the exercise program the performance must be supervised to prevent complications. The beginning is easiest, organized by ergometer exercise three times a week. During the beginning of the exercise program ECG is monitored until the individual response to exercise is known and steady state dosage in exercise is obtained. In long-term exercise training the exercise is best composed of such exercise forms increasing cardiovascular, metabolic, and muscular adaptation to the increased aerobic work stimulus. Thus the proper exercise froms include brisk walking, jogging, ergometer cycling, cycling, and swimming. Tennis playing could also be practiced without too active spurts. Still those exercise forms with easily predictable heart rate levels are preferable for persons with CHD. Occasional telemetric monitoring is useful in detecting possible ectopic heart beats induced by the exercise. Isometric exercises (weight lifting, body building) can not be recommended to persons with CHD as these exercise forms often increase disproportionally systolic blood pressure and lead to the high increase of the left ventricular strain. A regular diary, where all the exercises, drugs during the exercise, and maximal heart rates are marked is also recommended during a long-term training. In the long run after 3, 6, and 12 months an exercise stress test provides further information on the development of the physical fitness, stress resistance preceding symptoms, cardiac response to the exercise, and might lead to the new evaluation of the proper exercise load.

It is known that occasionally there are reports on sudden deaths in even healthy persons who are jogging. Even persons without any known disease have deceased while exercising. However, in large groups of persons with CHD, the incidence of sudden deaths and successfully resuscitated heart blocks have not increased in exercising persons in comparison with spontaneous heart blocks in unselected myocardial infarction populations.[110] A careful preexercise examination of CHD patients and good planning of the exercise program decreases the unexpected heart troubles in exercising persons with CHD.

VI. RESPONSE TO TRAINING

One of the most well-documented responses to training is the increase in the total work load in the exercise stress test. The maximal oxygen uptake increases by 15 to 25%.[95,96] The resting heart rate decreases and the heart rate response to increases in work load in the exercise stress test is lower than in untrained persons. The work load can be increased more after the exercise training than before in persons with CHD preceding the appearance of the chest pain, which might mean either the decreased oxygen demand of the heart or the increased coronary perfusion of the heart.[104] In postinfarction patients the exercise program also decreases the resting heart rate, and decreases the descent of the ST slope in ECG in exercise.[86,108,111] The oxygen uptake also improved.

In a recent study the exercise response in CHD persons with the decreased left ventricular function was studied.[112] This study included persons with a myocardial infarction 6 weeks or earlier from the examinations. All had the ejection fraction less than 40%. The presence of CHD was confirmed angiographically. These persons had in-

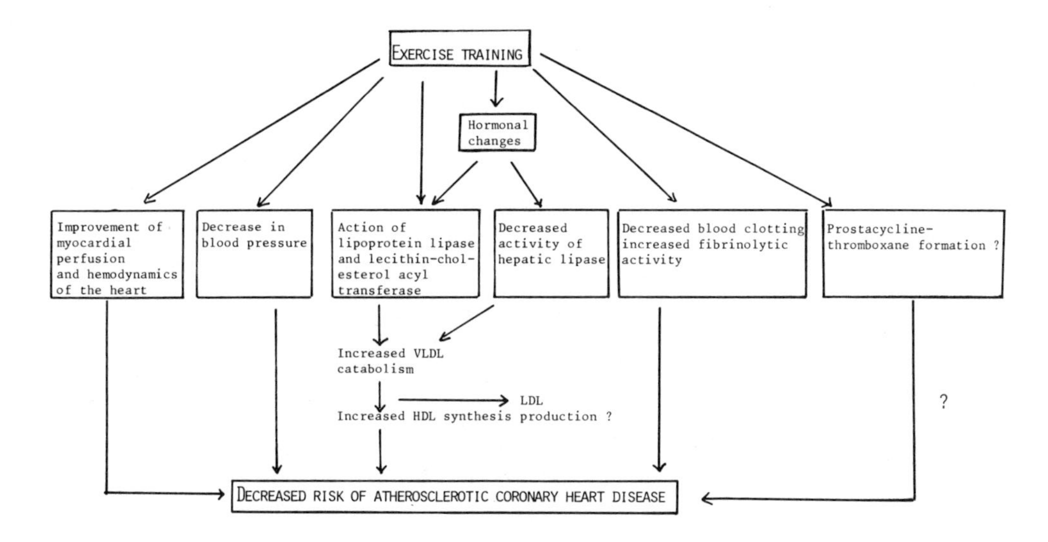

FIGURE 1. A schematic presentation on the possible pathways exercise training may intervene in the coronary heart disease.

dividual exercise training consisting of walking, jogging, and ergometer cycling at a gradually increasing heart rate up to 70 to 85% of the heart rate obtained in the exercise stress test. The program was continued under surveillance 6 to 8 weeks and thereafter the persons trained mainly unsupervised but had a diary. The maximal work load increased significantly, the resting heart rate decreased, and the oxygen uptake increased despite the unchanged ejection fraction during the course of the exercise program. Thus, even a decreased left ventricular function in CHD patients does not necessarily prevent exercise.

VII. CONCLUSIONS

Exercise training has positive effects on body functions and also on physical fitness both in healthy persons and in those having CHD or hyperlipemic metabolic disorders. Although no conclusive long-term follow-up studies exist to confirm the positive effects of exercise training, many experimental, epidemiological, cross-sectional, and longitudinal studies have proved valuable in collecting information on the effects of the physical activity. In the worst cases no effects have been found but no definitely negative effects of exercise have been confirmed and a large majority of the studies indicate positive effects of exercise on health and also probably help to delay or even prevent CHD.

Physical activity and consequently physical fitness mediate their effects via many pathways (Figure 1). The improvement of hemodynamics and the heart perfusion are advantageous in CHD as well as in healthy persons. Also, the possible decrease in blood pressure is advantageous. Other factors decreasing the risks for the coronary heart disease are increased fibrinolytic activity and hormonal changes. The physical activity has direct effects on the lipolytic activities and on the lecithin cholesterol acyltransferase which changes increase the HDL synthesis. Both the lipoprotein lipase and lecithin cholesterol acyltransferase activities increase due to the exercise while the hepatic lipase activity decreases. This yields increased catabolism of VLDL fraction due to the triglyceride hydrolysis, and consequently increased HDL production. Exercise training may also catalyze these reactions by causing changes in the hormonal levels as insulin, glucocorticoids, possibly sex hormones, and catecholamines, whose changes might regulate lipolytic enzymes. Thus, physical activity causing all these changes together may also increase the quality of life in patients with coronary heart disease.[113]

REFERENCES

1. **Froelicher, V. F.,** Does exercise conditioning delay progression of myocardial ischemia in coronary atherosclerotic heart disease, *Cardiovasc. Clin.,* 8, 11, 1977.
2. **Adelstein, A. M.,** Some aspects of cardiovascular mortality in South Africa, *Br. J. Prev. Soc. Med.,* 17, 29, 1963.
3. **Bergstrand, R., Vedin, A., Whilhelmsson, C., Wallin, J., Wedel, H., and Wilhelmsen, L.,** Myocardial infraction among men below age 40, *Br. Heart J.,* 40, 783, 1978.
4. **Blackburn, H., Taylor, H. L., and Keys, A.,** Coronary heart disease in seven countries, *Circulation,* 41(Suppl. 1), 1954, 1970.
5. **Breslow, L. and Buell, P.,** Mortality from coronary heart disease and physical activity of work in California, *J. Chronic Dis.,* 11, 421, 1960.
6. **Brunner, D.,** Myocardial infarction among members of communal settlements in Israel, *Lancet,* 2, 1049, 1960.
7. **Brunner, D.,** The influence of physical activity on incidence and prognosis of ischemic heart disease, in *Prevention of Ischemic Heart Disease,* Raab, W., Ed., Charles C Thomas, Springfield, Ill., 1966, 1.
8. **Cassel, J., Heyden, S., Bartel, A. G., Kaplan, B. H., Tyroler, H. A., Cornoni, J. C., and Hames, C. G.,** Occupation and physical activity and coronary heart disease, *Arch. Intern. Med.,* 128, 920, 1971.
9. **Chapman, J. M. and Massey, F. J.,** The interrelationship of serum cholesterol, hypertension, body weight, and risk of coronary disease, *J. Chronic Dis.,* 17, 933, 1964.
10. **Costas, R., Jr., Garcia-Palmieri, M. R., Nazario, E., and Sorlie, P. D.,** Relation of lipids, weight and physical activity to incidence of coronary heart disease: the Puerto Rico heart study, *Am. J. Cardiol.,* 42, 653, 1978.
11. **Ferguson, R. J., Petitclerc, R., Choquette, G., Chaniotis, L., Gauthier, P., Huot, R., Allard, C., Vankowski, L., and Campean, L.,** Effect of physical training on treadmill exercise capacity, collateral circulation and progression of coronary disease, *Am. J. Cardiol.,* 34, 765, 1974.
12. **Fletcher, G. F. and Cantwell, J. D.,** Outpatient gym exercise program for patients with recent myocardial infarction, *Arch. Intern. Med.,* 134, 63, 1974.
13. **Forssman, O. and Lindegard, B.,** The post-coronary patient. A multidisciplinary investigation of middle-aged Swedish males, *J. Psychosom. Res.,* 3, 89, 1958.
14. **Frank, C. W., Weinblatt, E., Shapiro, S., and Sager, R. V.,** Physical inactivity as a lethal factor in myocardial infarction among men, *Circulation,* 34, 1022, 1966.
15. **Gordon, T., Sorlie, P., and McNamara, P.,** Physical activity and coronary vulnerability: the Framingham study, *Cardiol. Dig.,* 6, 28, 1971.
16. **Gottheiner, V.,** Long-range strenuous sports training for cardiac reconditioning and rehabilitation, *Am. J. Cardiol.,* 22, 426, 1968.
17. **Hellerstein, H. K.,** Effects of an active physical reconditioning intervention program on the clinical course of coronary artery disease, *Mal. Cardiovasc.,* 10, 461, 1969.
18. **Kahn, H. A.,** The relationship of reported coronary heart disease mortality to physical activity of work, *Am. J. Public Health,* 53, 1058, 1963.
19. **Kannel, W. B.,** Habitual level of physical activity and risk of coronary heart disease: the Framingham Study, *Can. Med. Assoc. J.,* 96, 811, 1967.
20. **Kattus, A. A., Jorgensen, C. R., Worden, R. E., and Alvaro, A. B.,** ST segment depression with near-maximal exercise: its modification by physical conditioning, *Chest,* 62, 678, 1972.
21. **Kavanagh, T., Shephard, R. H., and Pandit, V.,** Marathon running after myocardial infarction, *JAMA,* 229, 1602, 1974.
22. **Kennedy, C. C., Spiekerman, R. E., Mankin, H. T., et al.,** One year graduated exercise program for men with angina pectoris, *Mayo Clin. Proc.,* 51, 231, 1976.
23. **Marmot, M. G., Rose, G., Shipley, M., and Hamilton, P. J. S.,** Employment grade and coronary heart disease in British Civil servants, *J. Epidemiol. Community Health,* 32, 244, 1978.

24. **McDonough, J. R., Hames, C. G., Stulb, S. C., and Garrison, G. E.,** Coronary heart disease among Negroes and Whites in Evans County, Georgia, *J. Chronic Dis.*, 18, 443, 1965.
25. **Mitrani, Y., Karplus, H., and Brunner, D.,** Coronary atherosclerosis in cases of traumatic death, in *Physical Activity and Aging,* Vol. 4, Brunner, D., Ed., University Park Press, Baltimore, 1970, 241.
26. **Morris, J. N.,** Epidemiology and cardiovascular disease of middle age. I and II, *Mod. Concepts Cardiovasc. Dis.*, 29, 625, 1960.
27. **Morris, J. N., Adam, C., Chave, S., Sirey, C., and Epstein, L.,** Vigorous exercise in leisuretime and the incidence of coronary heart disease, *Lancet,* 1, 333, 1973.
28. **Morris, J. N. and Crawford, M. D.,** Coronary heart disease and physical activity of work, *Br. Med. J.*, 2, 1485, 1958.
29. **Morris, J. N., Heady, J. A., Raffle, P. A., Roberts, C. G., and Parks, J. W.,** Coronary heart disease and physical activity of work, *Lancet*, 2, 1111, 1953.
30. **Morris, J. N., Heady, J. A., and Raffle, P. A.,** Physique of London busmen, *Lancet,* 2, 569, 1956.
31. **Morris, J. N., Kagan, A., Pattison, D. C., Gardner, M. J., and Raffle, P. A. B.,** Incidence and prediction of ischaemic heart disease in London busmen, *Lancet,* 2, 553, 1966.
32. **Oliver, R. M.,** Physique and serum lipids of young London busmen in relation to ischaemic heart disease, *Br. J. Intern. Med.*, 24, 181, 1967.
33. **Olsson, A. G., Kaijser, L., Walldins, G., Logan, R. L., Riermersam, R. A., and Oliver, M. F.,** Risk factors for ischaemic heart disease, with emphasis on nutrition and exercise, *Bibl. Nutr. Dieta,* 27, 18, 1979.
34. **Paffenbarger, R. S. and Hale, W. E.,** Work activity and coronary heart mortality, *N. Engl. J. Med.*, 292, 545, 1975.
35. **Paffenbarger, R. S., Jr., Hale, W. E., Brand, R. J., and Hyde, R. T.,** Work-energy level, personal characteristics, and fatal heart attack: a birthcohort effect, *Am. J. Epidemiol.*, 105, 200, 1977.
36. **Paffenbarger, R. S., Laughlin, M. E., Gima, A. S., and Black, R. A.,** Work activity of longshoremen as related to death from coronary heart disease and stroke, *N. Engl. J. Med.*, 282, 1109, 1970.
37. **Paffenbarger, R. A., Jr., Wing, A. L., and Hyde, R. T.,** Physical activity as an index of heart attack risk in college alumni, *Am. J. Epidemiol.*, 108, 161, 1978.
38. **Paul, O.,** Physical activity and coronary heart disease. II, *Am. J. Cardiol.*, 23, 303, 1969.
39. **Pesonen, E., Norio, R., and Sarna, S.,** Thickenings in the coronary arteries in infancy as an indication of genetic factors in coronary heart disease, *Circulation,* 51, 218, 1975.
40. **Rechnitzer, P. A., Pickard, H. A., Paivio, A. U., Yuhasz, M. S., and Cunningham, D.,** Long-term followup study of survival and recurrence rates following myocardial infarction in exercising and control subjects, *Circulation,* 45, 853, 1972.
41. **Redwood, D. R., Rosing, D. R., and Epstein, S. E.,** Circulatory and symptomatic effects of physical training in patients with coronary artery disease and angina pectoris, *N. Engl. J. Med.*, 286, 959, 1972.
42. **Rose, G., Prineas, R. J., and Mitchell, J. R.,** Myocardial infarction and the intrinsic calibre of coronary arteries, *Br. Heart, J.*, 29, 548, 1967.
43. **Roseman, R. H.,** The influence of different exercise patterns on the incidence of coronary heart disease in the western collaborative group study, in *Physical Activity and Aging,* Vol. 4, Brunner, D., Ed., University Park Press, Baltimore, 1970, 267.
44. **Roseman, R. H., Brand, R. J., Jenkins, C. D., Friedman, M., Straus, R., and Wurm, M.,** Coronary heart disease in the western collaborative group study, *JAMA*, 233, 872, 1975.
45. **Shanoff, H. M. and Little, J. A.,** Studies of male survivors of myocardial infarction to "essential" atherosclerosis. I. Characteristics of the patients, *Can. Med. Assoc. J.*, 84, 519, 1961.
46. **Spain, D. M. and Bradess, V. A.,** Occupational physical activity and the degree of coronary atherosclerosis in "normal" men, *Circulation,* 22, 239, 1960.
47. **Stamler, J., Berkson, D. M., Lindberg, H. A., Whipple, I. T., Miller, W., Mojonnier, L., Hall, Y. F., Soyugenc, R., and Levinson, M. J.,** Long-term epidemiologic studies on the possible role of physical activity and physical fitness in the prevention of premature clinical coronary heart disease, in *Physical Activity and Aging,* Vol. 4, Brunner, D., Ed., University Park Press, Baltimore, 1970, 274.
48. **Stamler, J., Kjelsberg, M., and Hall, Y.,** Epidemiologic studies on cardiovascular-renal diseases. I. Analysis of mortality by age-race-sex-occupation, *J. Chronic. Dis.*, 12, 440, 1960.
49. **Stamler, J., Lindberg, H. A., Berkson, D. M., Shaffer, A., Miller, M., and Poindexter, A.,** Prevalence and incidence of coronary heart disease in strata of the labor force of a Chicago industrial corporation, *J. Chronic. Dis.*, 11, 405, 1960.

50. **Taylor, H. L., Klepetar, E., Keys, A., Parlin, W., Blackburn, H., and Puchner, T.,** Death rates among physically active and sedentary employees of the railroad industry, *Am. J. Public Health,* 52, 1697, 1962.
51. **Tibblin, G., Wilhelmsen, L., and Werko, L.,** Risk factors for myocardial infarction and death due to ischemic heart disease and other causes, *Am. J. Cardiol.,* 35, 14, 1975.
52. **Werko, L.,** Can we prevent heart disease?, *Ann. Intern. Med.,* 74, 278, 1971.
53. **Wilhelmsen, L., Sanne, H., Elmfeldt, D., Grimby, G., Tibblin, G., and Wedel, H.,** A controlled trial of physical training after myocardial infarction, *Prevent. Med.,* 4, 491, 1975.
54. **Zukel, W. J., Lewis, R., Enterline, P., Painter, R. C., Ralston, L. S., Fawcett, R. M., Meredith, A. P., and Peterson, B.,** A short-term community study of the epidemiology of coronary heart disease, *Am. J. Public Health,* 49, 1630, 1959.
55. **Castelli, W. P., Doyle, J. T., Gordon, T., Hames, C. G., Hjortland, M. C., Hulley, S. B., Kagan, A., and Zukel, W. J.,** HDL cholesterol and other lipids in coronary heart disease. The cooperative lipoprotein phenotyping study, *Circulation,* 55, 767, 1977.
56. **Kannel, W. B.,** Hypertension, blood lipids, and cigarette smoking as co-risk factors for coronary heart disease, *N. Y. Acad. Sci.,* 304, 128, 1978.
57. **Jenkins, P. J., Harper, R. W., and Nestel, P. J.,** Severity of coronary atherosclerosis related to lipoprotein concentration, *Br. Med. J.,* 281, 388, 1978.
58. **Albers, J. J., Cheung, M. C., and Hazzard, W. R.,** High-density lipoproteins in myocardial infarction survivors, *Metabolism,* 27, 479, 1978.
59. **Erkelens, D. W., Albers, J. J., Hazzard, W. R., Frederick, R. C., and Bierman, E. L.,** High-density lipoprotein-cholesterol in survivors of myocardial infarction, *JAMA,* 242, 2185, 1979.
60. **Pometta, D., Micheli, H., Jornot, C., and Scherrer, J. R.,** HDL-Cholesterol abaisse chez les proches parents et les malades victimes d'infarctus du myocarde, *Schweiz. Med. Wschr.,* 108, 1888, 1978.
61. **Kaukola, S., Manninen, V., and Halonen, P. I.,** Serum lipids with special reference to HDL cholesterol and triglycerides in young male survivors of acute myocardial infarction, *Acta Med. Scand.,* 208, 41, 1980.
62. **Tan, M. H., Macintosh, W., Weidon, K. L., Kapoor, A., Chandler, B. M., and Hindmarsh, T. J.,** Serum high density lipoprotein cholesterol in patients with abnormal coronary arteries, *Atherosclerosis,* 37, 187, 1980.
63. **Lewis, B., Carlson, L. A., Mancini, M., and Micheli, H.,** Serum lipoprotein abnormalities in ischaemic heart disease, *Postgrad. Med. J.,* 51, 37, 1975.
64. **Wilson, P. W., Garrison, R. J., Castelli, W. P., Feinleib, M., McNamara, P. M., and Kannel, W. P.,** Prevalence of coronary heart disease in the Framingham offspring study: role of lipoprotein cholesterols, *Am. J. CArdiol.,* 46, 649, 1980.
65. **Onitiri, A. C. and Jover, E.,** Comparative serum apolipoprotein studies in ischaemic heart disease and control subjects, *Clin. Chim. Acta,* 108, 25, 1980.
66. **Barboriak, J. J., Anderson, A. J., Rimm, A. A., and King, J. F.,** High density lipoprotein cholesterol and coronary artery occlusion, *Metabolism,* 28, 735, 1979.
67. **Pearson, T. A., Bulkley, B. H., Achuff, S. C., Kwiterovich, P. O., and Gordis, L.,** The association of low levels of HDL cholesterol and arteriographically defined coronary artery disease, *Am. J. Epidemiol.,* 109, 285, 1979.
68. **Wieland, H., Seidel, D., Wiegand, V., and Kreuzer, H.,** Serum lipoproteins and coronary artery disease (CAD). Comparison of the lipoprotein profile with the results of coronary angiography, *Atherosclerosis,* 36, 269, 1980.
69. **Einfeldt, H., Milz, H., and Wimmer, P. P.,** Möglichkeiten zur aktiven Prävention von Gefässerkrankungen, *Münch. Med. Wschr.,* 121, 387, 1979.
70. **Enger, S. Chr., Herbjørnsen, K., Erikssen, J. C., and Fretland, A.,** High density lipoproteins (HDL) and physical activity: the influence of physical exercise, age and smoking on HDL-cholesterol and the HDL-/total cholesterol ratio, *Scand. J. Clin. Lab. Invest.,* 37, 251, 1977.
71. **Garman, J. F.,** Coronary risk factor intervention—a review of physical activity and serum lipids, *Am. Corr. Ther. J.,* 32, 183, 1978.
72. **Hartung, G. H., Foreyt, J. P., Mitchell, R. E., Vlasek, I., and Gotto, A. M., Jr.,** Relation of diet to high-density-lipoprotein cholesterol in middleaged marathon runners, joggers, and inactive men, *N. Engl. J. Med.,* 302, 357, 1980.
73. **Huttunen, J., Länsimies, E., Voutilainen, E., Ehnholm, Ch., Hietnaen, E., Penttilä, I., Siitonen, O., and Rauramaa, R.,** Effect of moderate physical exercise on serum lipoproteins. A controlled clinical trial with special reference to serum high-density lipoproteins, *Circulation,* 60, 1220, 1979.
74. **Lehtonen, A. and Viikari, J.,** Serum triglycerides and cholesterol and serum high-density lipoprotein cholesterol in highly physically active men, *Acta Med. Scand.,* 204, 111, 1978.

75. **Nikkilä, E. A., Taskinen, M. R., Rehunen, S., and Härkönen, M.,** Lipoprotein lipase activity in adipose tissue and skeletal muscle of runners: relation to serum lipoproteins, *Metabolism,* 27, 1661, 1978.
76. **Peltonen, P., Marniemi, J., Hietanen, E., Vuori, I., and Ehnholm, C.,** Changes in serum lipids, lipoproteins and heparin releasable lipolytic enzymes during moderate physical training in man. A longitudinal study, *Metabolism,* 30, 518, 1981.
77. **Schwane, J. A. and Cundiff, D. E.,** Relationships among cardiorespiratory fitness, regular physical activity and plasma lipids in young adults, *Metabolism,* 28, 771, 1979.
78. **Schlierf, G.,** Arteriosklerose-Möglichkeiten für Prophylaxe und Therapie, *Internist,* 19, 632, 1978.
79. **Wood, P. D. and Haskell, W. L.,** The effect of exercise on plasma high density lipoproteins, *Lipids,* 14, 417, 1979.
80. **Berger, G. M. B.,** High-density lipoproteins in the prevention of atherosclerotic heart disease. I. Epidemiological and family studies, *S. Afr. Med. J.,* 54, 689, 1978.
81. **Galyean, J. R.,** Risk factors for coronary heart disease, *Southern Med. J.,* 71, 684, 1978.
82. **Imai, Y., Shino, A., Asano, T., Matsumura, H., and Kakinuma, A.,** Increase of serum high density lipoprotein with progression and regression of aortic lipid deposition in rats, *Atherosclerosis,* 34, 329, 1979.
83. **Wissler, R. W.,** Progression and regression of atherosclerotic lesions, *Adv. Exp. Med. Biol.,* 104, 77, 1978.
84. **Blackburn, H.,** Coronary disease prevention. Controversy and professional attitudes, *Adv. Cardiol.,* 20, 10, 1977.
85. **Blackburn, H.,** Concepts and controversies about the prevention of coronary heart disease, *Conn. Med.,* 41, 7, 1977.
86. **Hartung, G. H.,** Physical activity and coronary heart disease risk—a review, *Am. Corr. Ther. J.,* 31, 110, 1977.
87. **Hulley, S. B., Cohen, R., and Widdowson, G.,** Plasma high-density lipoprotein cholesterol level. Influence of risk factor intervention, *JAMA,* 238, 2269, 1977.
88. **Mann, G. V., Garrett, H. L., Farhi, A., Murray, H., Billings, F. T., Shute, E., and Schwarten, S. E.,** Exercise to prevent coronary heart disease. An experimental study of the effects of training on risk factors for coronary disease in men, *Am. J. Med.,* 46, 12, 1969.
89. **Margolis, S.,** Physician strategies for the prevention of coronary heart disease, *Johns Hopkins Med. J.,* 141, 170, 1977.
90. **Scheele, K., Herzog, W., Ritthaler, G., Wirth, A., and Weicker, H.,** Metabolic adaptation to prolonged exercise, *Eur. J. Appl. Physiol.,* 41, 101, 1979.
91. **Simonelli, C. and Eaton, R. P.,** Cardiovascular and metabolic effects of exercise. The strong case for conditioning, *Postgrad. Med.,* 63, 71, 1978.
92. **Lehtonen, A. and Viikari, J.,** The effect of vigorous physical activity at work on serum lipids with a special reference to serum high-density lipoprotein cholesterol, *Acta Physiol. Scand.,* 104, 117, 1978.
93. **Vodak, P. A., Wood, P. D., Haskell, W. L., and Williams, P. T.,** HDL-Cholesterol and other plasma lipid and lipoprotein concentrations in middle-aged male and female tennis players, *Metabolism,* 29, 745, 980.
94. **Boyer, J. L. and Kasch, F. W.,** Exercise therapy in hypertensive men, *JAMA,* 211, 1668, 1970.
95. **Fox, E. L.,** Methods and effects of physical training, *Ped. Ann.,* 7, 690, 1978.
96. **Hanson, P. G., Giese, M. D., and Corliss, R. J.,** Clinical guidelines for exercise training, *Postgrad. Med.,* 67, 120, 1980.
97. **Pollock, M. L., Cureton, T. K., and Greninger, L.,** Effects of frequency of training on working capacity, cardiovascular function, and body composition of adult men, *Med. Sci. Sports,* 1, 70, 1969.
98. **Sedgwick, A., Brotherhood, J. R., Harris-Davidson, A., Taplin, R. E., and Thomas, D. W.,** Long-term effects of physical training programme on risk factors for coronary heart disease in otherwise sedentary men, *Br. Med. J.,* 281, 7, 1980.
99. **Bonanno, J. A.,** Coronary risk factor modification by chronic physical exercise, in *Exercise in Cardiovascular Health and Disease,* Amsterdam, E. A., Wilmore, J. H., and deMaria, A. N., Eds., Yorke Medical Books, New York 1977, 274.
100. **Lee, G., Amsterdam, E. A., deMaria, A. N., Davis, G., LaFave, T., and Mason, D. T.,** Effect of exercise on hemostatic mechanisms, in *Exercise in Cardiovascular Health and Disease,* Amsterdam, E. A., Wilmore, J. H., and deMaria, A. N., Eds., Yorke Medical Books, New York, 1977, 122.
101. **Olson, R. E.,** Is there an optimum diet for the prevention of coronary heart disease?, in *Nutrition, Lipids, and Coronary Heart Disease,* Lewy, R., Rifkind, B., Dennis, B., and Ernst, N., eds., Raven Press, New York, 1979, 349.

102. **Neill, W. A. and Oxendine, J. M.,** Exercise can promote coronary collateral development without improving perfusion of ischemic myocardium, *Circulation,* 60, 1513, 1979.
103. **Margolis, S.,** Physician strategies for the prevention of coronary heart disease, *John Hopkins Med. J.,* 141, 170, 1977.
104. **Neill, W. A.,** Coronary and systemic circulatory adaptations to exercise training and their effects on angina pectoris, in *Exercise in Cardiovascular Health and Disease,* Amsterdam, E. A., Wilmore, J. H., and deMaria, A. N., Eds., Yorke Medical Books, New York, 1977, 137.
105. **Haskell, W. L. and DeBusk, R.,** Cardiovascular responses to repeated treadmill exercise testing soon after myocardial infarction, *Circulation,* 60, 1247, 1979.
106. **Sami, M., Kraemer, H., and DeBusk, R. F.,** The prognostic significance of serial exercise testing after myocardial infarction, *Circulation,* 60, 1238, 1979.
107. **Théroux, P., Waters, D. D., Halphen, C., Debaisieux, J.-C., and Mizgala, H. F.,** Prognostic value of exercise testing soon after myocardial infarction, *N. Engl. J. Med.,* 301, 341, 1979.
108. **DeBusk, R. F., Houston, N., Haskell, W., Fry, G., and Parker, M.,** Exercise training soon after myocardial infarction, *Am. J. Cardiol.,* 44, 1223, 1979.
109. **Streja, D. and Mymim, D.,** Moderate exercise and high-density lipoprotein cholesterol. Observations during a cardiac rehabilitation program, *JAMA,* 242, 2190, 1979.
110. **Haskell, W. L.,** Cardiovascular complications during exercise training of cardiac patients, *Circulation,* 59, 549, 1978.
111. **Haskell, W. L.,** Physical following myocardial infarction, in *Exercise in Cardiovascular Health and Disease,* Amsterdam, E. A., Wilmore, J. H., and deMaria, A. N., Eds., Yorke Medical Books, New York, 1977, 344.
112. **Lee, A. P., Ice, R., Blessey, R., and Sanmarco, M.,** Long-term effects of physical training on coronary patients with impaired ventricular function, *Circulation,* 50, 1519, 1979.
113. **Naughton, J.,** Cardiac rehabilitation: principles, techniques, applications, in *Exercise in Cardiovascular Health and Disease,* Amsterdam, E. A., Wilmore, J. H., and deMaria, A. N., Eds., Yorke Medical Books, New York, 1977, 364.

Index

INDEX

A

B

C

D

E

F

G

H

I

J

L

M

N

O

P

U

V

W

X

Y

Z